Springer Undergraduate Mathematics Series

Springer-Verlag London Ltd.

Advisory Board

Professor P.J. Cameron *Queen Mary and Westfield College*
Dr M.A.J. Chaplain *University of Dundee*
Dr K. Erdmann *Oxford University*
Professor L.C.G. Rogers *University of Bath*
Dr E. Süli *Oxford University*
Professor J.F. Toland *University of Bath*

Other books in this series

Ioan James

Topologies and Uniformities

With 25 Figures

 Springer

Ioan Mackenzie James, MA, DPhil, FRS
Mathematical Institute, Oxford University, 24–29 St Giles, Oxford OX1 3LB, UK

Cover illustration elements reproduced by kind permission of:
Aptech Systems, Inc., Publishers of the GAUSS Mathematical and Statistical System, 23804 S.E. Kent-Kangley Road, Maple Valley, WA 98038, USA. Tel: (206) 432 - 7855 Fax (206) 432 - 7832 email: info@aptech.com.URL: www.aptech.com
American Statistical Association: Chance Vol 8 No 1, 1995 article by KS and KW Heiner 'Tree Rings of the Northern Shawangunks' page 32 fig 2
Springer-Verlag: Mathematica in Education and Research Vol 4 Issue 3 1995 article by Roman E Maeder, Beatrice Amrhein and Oliver Gloor 'Illustrated Mathematics: Visualization of Mathematical Objects' page 9 fig 11, originally published as a CD ROM 'Illustrated Mathematics' by TELOS: ISBN 0-387-14222-3, german edition by Birkhauser: ISBN 3-7643-5100-4.
Mathematics in Education and Research Vol 4 Issue 3 1995 article by Richard J Gaylord and Kazume Nishidate 'Traffic Engineering with Cellular Automata' page 35 fig 2. Mathematica in Education and Research Vol 5 Issue 2 1996 article by Michael Trott 'The Implicitization of a Trefoil Knot' page 14.
Mathematics in Education and Research Vol 5 Issue 2 1996 article by Lee de Cola 'Coins, Trees, Bars and Bells: Simulation of the Binomial Process page 19 fig 3. Mathematica in Education and Research Vol 5 Issue 2 1996 article by Richard Gaylord and Kazume Nashidate 'Contagious Spreading' page 33 fig 1. Mathematica in Education and Research Vol 5 Issue 2 1996 article by Joe Buhler and Stan Wagon 'Secrets of the Madelung Constant' page 50 fig 1.

ISBN 978-1-85233-061-3

British Library Cataloguing in Publication Data
James, Ioan M.
   Topologies and uniformities
   1. Topology   2. Topological spaces
   I. Title
   514
   ISBN 978-1-85233-061-3   ISBN 978-1-4471-3994-2 (eBook)
   DOI 10.1007/978-1-4471-3994-2

Library of Congress Cataloging-in-Publication Data
James, I. M. (Ioan Mackenzie), 1928–
   Topologies and uniformities/I. M. James.
      p.   cm.
   Includes bibliographical refrerences and index.
   ISBN 978-1-85233-061-3
   1. Topology.   2. Topological spaces.   3. Uniform spaces.
I. Title.
QA611.J337   1999   98-41748
514–dc21   CIP

Expanded and revised version of *Topological and Uniform Spaces*, published by Springer-Verlag London Ltd.

Typesetting by BC Typesetting, Bristol BS31 1NZ

12/3830-543210 Printed on acid-free paper

Contents

Introduction

This book is based on lectures I have given to senior undergraduate and graduate audiences at Oxford and elsewhere over the years. My aim has been to provide an outline of both the topological theory and the uniform theory, with an emphasis on the relation between the two. Although I hope that the prospective specialist may find it useful as an introduction it is the non-specialist I have had more in mind in selecting the contents. Thus I have tended to avoid the ingenious examples and counterexamples which often occupy much of the space in books on general topology, and I have tried to keep the number of definitions down to the essential minimum. There are no particular prerequisites but I have worked on the assumption that a potential reader will already have had some experience of working with sets and functions and will also be familiar with the basic concepts of algebra and analysis.

An earlier version of the present book appeared in 1987 under the title *Topological and Uniform Spaces*. When the time came for a new edition I came to the conclusion that, rather than just making the necessary corrections, it would be better to make more substantial alterations. Parts of the text have been rewritten and new material, including new diagrams, added. The sets of exercises at the end of each chapter have been revised and worked solutions to all of them provided at the end of the book. Also a historical note has been added and the Select Bibliography has been expanded. Altogether these changes seemed sufficient to justify a new title.

The book divides naturally into three sections. Thus the first six chapters are devoted to the topological theory while the next two are devoted to the uniform theory. The last four, which are independent of each other, draw on ideas from both the topological section and the uniform section.

After a few preliminaries, in which notation and terminology are established, the first chapter of the topological section mainly deals with the basic axioms. Illustrations are taken from interval topologies and metric topologies, with special reference to the real line. No previous knowledge of metric spaces is assumed. An outline of the theory of filters is included.

The second chapter is concerned with continuity: topology is about continuous functions just as much as topological spaces. The topological product is dealt with here. Also topological groups are introduced at this stage both because of their intrinsic interest and because they provide such excellent illustrations of points in the general theory. Subspaces and quotient spaces are considered in Chapter 3, with a wide range of examples.

Most accounts of the theory go on to discuss separation axioms, connectedness and so forth at this point. But in my view compactness should come first, because of its fundamental importance. I believe the concept arises most naturally from a discussion of open functions and closed functions. This is not the orthodox approach, of course, but I have tried to justify it by showing that all the usual properties of compact spaces such as the Heine–Borel theorem can be proved quite simply and directly from this approach. I also show how compactness can be characterized in terms of filters, and incidentally show how the best-known characterization of compactness, in terms of open coverings, can be obtained. The general Tychonoff theorem is proved, followed by some observations on the subject of function spaces. This material occupies Chapters 4 and 5.

Chapter 6 is devoted to the separation axioms: the basic properties of Hausdorff, regular and normal spaces are established. In a later chapter there is an account of the corresponding functional separation axioms.

Chapter 7 of the uniform section deals with the basic axioms of uniform structure, with illustrations from topological groups and metric spaces. I have tried to show how the idea of a uniformity is a very natural one, in many ways more natural than the idea of a topology. This leads on to the notion of uniform continuity: the uniform theory is about uniformly continuous functions just as much as uniform spaces. I also deal with the uniform product with subspaces and, to a limited extent, with quotient spaces.

In Chapter 8 the connection between the uniform and the topological theories is established. It becomes clear at this stage that results about topological groups and metric spaces found earlier can be regarded as special cases of results about uniform spaces. The chapter continues with a discussion of the Cauchy condition, both for sequences and for filters. This lays the foundation for a subsequent chapter on completeness and completion.

The first of the last four chapters is concerned with connectedness. I show how connectedness, local connectedness and pathwise-connectedness are defined, for a topological space. I also discuss connectedness and local connectedness for uniform spaces.

The second of the last four chapters is concerned with countability. The first and second axioms are discussed, countable compactness and sequential compactness are considered, also the Lindelöf property and separability. In the next chapter we return to the separation axioms. After a glance at the functional Hausdorff property we discuss functional (i.e. complete) regularity. In particular, we show that complete regularity is a necessary and sufficient condition for a topological space to be uniformixable. We also prove Urysohn's theorem, to the effect that normal spaces are functionally normal.

The final chapter is concerned with completeness and completion, both for metric spaces and for uniform spaces. Metric completions are constructed both using the space of bounded continuous real-valued functions and via equivalence classes of Cauchy sequences. Then the uniform completion is constructed via equivalence classes of Cauchy filters.

Among those who made valuable suggestions about the original work were Dr Alan Pears and Dr Wilson Sutherland. I would like to thank them again and also to thank Dr Ian Stares who not only provided the worked solutions to the exercises but also made some helpful comments about the main text.

Ioan James
Mathematical Institute
University of Oxford

Historical Note

Progress in mathematical research is usually a gradual process until the point is reached where someone (or often several people independently) sees just what is needed to achieve a satisfactory theory and open up a whole new mathematical landscape. For example, take the real number system. The real line is a concept which everyone can grasp intuitively. However in the nineteenth century mathematicians began to feel the need for greater rigour in the field of analysis, for which a first requirement was a more satisfactory definition of the real number system. This was provided by Cantor and by Dedekind in 1872. One used Cauchy sequences of the rationals, the other used sections of the rationals. They arrived, quite independently, at equally good but quite different ways of defining the real numbers, and in doing so they were, of course, building on earlier work.

However, the quest for greater rigour did not stop there. One of the requirements of analysis was the need for a satisfactory theory of convergence. In the first part of the nineteenth century this had been considered by Bolzano and Cauchy, amongst others, in what has been described as a "first revolution" of rigour. Later there was a "second revolution", led by Weierstrass, which was based on the work of Cantor and Dedekind.

The concept of function evolved gradually as the century progressed. At first it meant a specific formula, such as a polynomial, but later it came to mean what it does today. The concept of continuity evolved at the same time. In a very significant development Volterra (1883) began to treat classes of functions, rather than individual functions, and the modern theory of functional analysis developed from this.

Until the mid-nineteenth century the concept of space was limited to euclidean spaces and subspaces of euclidean spaces. However Riemann and others began to conceive of spaces in a more general sense. After the theory of sets was established by Cantor and others, it was possible to say, at least tentatively, what was meant by the term space. It was a set with some additional structure such that the concept of limit, for example, could be defined just as it was in the case of euclidean space. Thus suppose that, following Volterra, we wish to consider the set of all continuous real-valued functions of a real variable, what is the appropriate additional structure which would enable us to discuss limits in such a set, which is certainly not a euclidean space?

By the turn of the century leading mathematicians such as Hilbert and Weyl were addressing the question of what this additional structure should be. The term "neighbourhood" began to be used which in the euclidean case meant all the points of the space within a certain distance of a given point. Finally it was Fréchet (1906) who introduced the concept of metric space, as we know it today, and developed the theory over the next few years.

In Fréchet's theory the convergence of sequences is fundamental. That was soon seen as too restrictive and the search for the perfect concept continued. Although others may have had much the same idea it was Hausdorff (1914) who laid down the set of axioms for what we would now call a Hausdorff space and developed a substantial part of the theory we know today. Hausdorff took systems of neighbourhoods as fundamental, in his theory. The first three of his four axioms define what we now call a topology, the last axiom is the one we now call the Hausdorff property.

So at last the notion of topological space emerged into the light of day, with that of metric space as an important special case, and it was possible to see that many of the concepts which had been developed over the years in relation to less general spaces could be understood much better in this more general framework, for example the concept of compactness, which had begun to appear in special cases almost a century earlier.

At first topology was regarded as a branch of set theory but as its full potential became apparent it became increasingly regarded as a subject in its own right. Hausdorff's version of the theory was followed by alternative (but equivalent) versions, such as that of Kuratowski (1922), who took the notion of closure as fundamental, of Alexandroff (1925) who took the family of open sets as fundamental, and of Sierpinski (1927), who developed the dual approach using closed rather than open sets. Although Fréchet's theory was superseded by that of Hausdorff we owe to him not only the abstract concept of metric space but also such other basic concepts as completeness, separability and sequential compactness. Later it was seen that a modification of Fréchet's approach, in which filters or directed sets replaced sequences, would overcome the defects caused by the countable nature of the terms of a sequence and have other advantages. Filters, under one name or another, were considered by several mathematicians earlier in the century but the theory was developed systematically by H. Cartan (1937).

Although the theory of Lie groups (topological groups where the underlying space is a manifold and the group operations are smooth) goes back to around 1870 the theory of topological groups in a more general sense seems not to have been considered until 1925 when Leja and Schreier, independently, made the basic definitions, rather in the spirit of Fréchet, and since then this too has developed into a major branch of modern mathematics.

It was Weil (1937) who wrote the first definitive study of uniform spaces and applied the theory to both metric spaces and topological groups. However the basic idea was already emerging early in the century, indeed the concepts of uniform continuity and uniform convergence were well understood by Weierstrass and by Cauchy before him.

The full history of the development of the mathematics in this book has yet to be written but there are a number of articles on particular topics in the literature which have assisted me in the writing of this historical note. Much of the research on topological and uniform spaces which has been carried out in the latter part of the twentieth century is of a technical nature, chiefly of interest to specialists. The theories presented in this book were mainly developed in the earlier part of the century but since then there has been important progress in the understanding of the basic ideas and my aim has been to take advantage of this rather than describe the fruits of the latest research.

1
Topological Spaces

1.1 General remarks

For the convenience of the reader I begin by collecting together some of the terminology and notation we shall be using later.

Following standard practice in topology I will generally refer to elements of sets as points, in the course of this work, but I will retain the term element where groups are concerned specifically.

If every point of the set X is also a point of the set Y I describe X as a *subset* of Y or Y as a *superset* of X, and write $X \subset Y$ or $Y \supset X$. If it is necessary to exclude the possibility that $X = Y$ I describe X as a *proper* subset of Y.

In dealing with subsets of a given set I say that a pair of subsets *intersect* if their intersection is non-empty, otherwise I say that they do not intersect or are *disjoint*. Strictly speaking, families of subsets should always be indexed but it is seldom necessary to mention the indexing set explicitly.

Let $\phi\colon X \to Y$ be a function, where X and Y are sets; we refer to X as the *domain* and to Y as the *codomain* of ϕ. We say that ϕ is *injective*, or is an *injection*, if for each point y of Y there exists at most one point x of X such that $\phi(x) = y$. We say that ϕ is *surjective*, or is a *surjection*, if for each point y of Y there exists at least one point x of X such that $\phi(x) = y$. We say that ϕ is *bijective*, or is a *bijection*, if there exists a function $\psi\colon Y \to X$ such that $\psi\phi$ is the identity on X and $\phi\psi$ is the identity on Y. Note that ϕ is bijective if and only if ϕ is both injective and surjective.

1.2 Direct and inverse images

Again let $\phi\colon X \to Y$ be a function, where X and Y are sets. The *direct image* of a subset H of X, with respect to ϕ, is the subset ϕH of Y consisting of all points y of Y such that $y = \phi(x)$ for some point x of H. The *inverse image* of a subset K of Y, with respect to ϕ, is the subset $\phi^{-1}K$ of X consisting of all points x of X such that $\phi(x) \in K$. The inverse image behaves well with respect to complementation, in that

$$\phi^{-1}(Y - K) = X - \phi^{-1}K.$$

In the case of the direct image all we can say is that $\phi X - \phi H$ is a subset of $\phi(X - H)$, in general a proper subset.

For each subset H of X the *saturation* $\phi^{-1}\phi H$ of H with respect to ϕ contains H as a subset, in general a proper subset, as illustrated on the left of Figure 1.1. When $H = \phi^{-1}\phi H$ we say that H is *saturated*. All subsets H are saturated when ϕ is injective.

For each subset K or Y the *cosaturation* $\phi\phi^{-1}K$ of K with respect to ϕ is contained in K as a subset, in general a proper subset, as illustrated on the right of Figure 1.1. When $K = \phi\phi^{-1}K$ we say that K is *cosaturated*. All subsets K are cosaturated when ϕ is surjective.

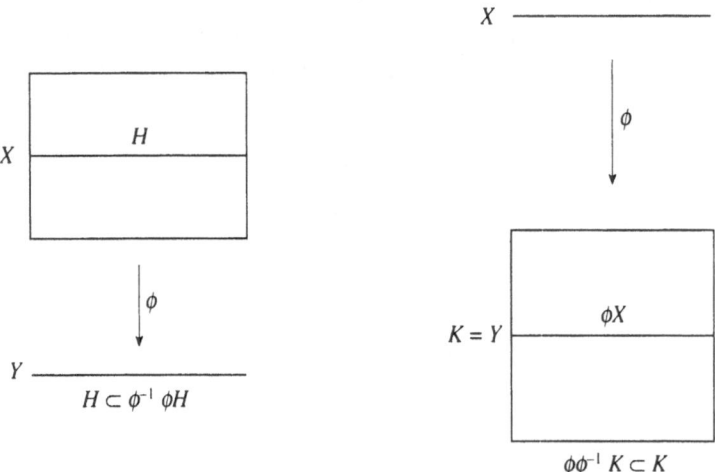

Figure 1.1.

Next we recall the behaviour of the direct image and the inverse image with respect to the operations of union and intersection. First let $\{H_j\}$ be a family of subsets of the set X. We have that

$$\phi\left(\bigcup H_j\right) = \bigcup (\phi H_j).$$

We also have that $\phi(\bigcap H_j)$ is a subset of $\bigcap(\phi H_j)$, in general a proper subset. Equality holds whenever ϕ is injective or, more generally, when not more than one of the sets H_j is non-saturated. Then let $\{K_j\}$ be a family of subsets of the set Y. We have that

$$\phi^{-1}(\bigcup K_j) = \bigcup (\phi^{-1} K_j),$$

and we also have that

$$\phi^{-1}(\bigcap K_j) = \bigcap (\phi^{-1} K_j).$$

Thus the inverse image behaves well with respect to the operations of complementation, union and intersection, but the direct image only behaves well with respect to the operation of union. It is important to remember these facts.

1.3 Cartesian products

Consider a family $\{X_j : j \in J\}$ of sets, where J is some indexing set. The cartesian product of the members of the family is a set $\prod X_j$ together with a family of functions

$$\pi_j : \prod X_j \to X_j \qquad (j \in J),$$

with the following characteristic property: for each set A the functions

$$\phi : A \to \prod X_j$$

correspond precisely to the families of functions

$$\phi_j = \pi_j \phi : A \to X_i \qquad (j \in J).$$

We refer to π_j as the jth *projection* of the cartesian product, and to ϕ_j as the jth *component* of the function ϕ.

   In case each of the sets X_j is the same set X, so that we are dealing with J copies of X, the cartesian product may be called the cartesian Jth power of X and written X^J. Then a function

$$\Delta : X \to X^J$$

is defined, called the *diagonal*, such that the jth component of Δ is the identity function on X, for each index j. There is, of course, no difference in logic between points ξ of X^J and functions $\xi : J \to X$. So $\pi_j(\xi)$ and $\xi(j)$ are just two ways of writing the same point of X; this change of notation will come up from time to time in what we are going to do and it is important not to be confused by it.

Given a family of functions

$$\phi_j \colon X_j \to Y_j \qquad (j \in J),$$

where X_j and Y_j are sets, there is a function

$$\prod \phi_j \colon \prod X_j \to \prod Y_j$$

such that the jth component of $\prod \phi_j$ is the composition $\phi_j \circ \pi_j$, for each index j. We refer to $\prod \phi_j$ as the product of the members of the family $\{\phi_j\}$. Note that if

$$\phi_j \colon X_j \to Y_j, \qquad \psi_j \colon Y_j \to Z_j \qquad (j \in J)$$

are families of functions, where X_j, Y_j and Z_j are sets, then

$$\prod(\psi_j \circ \phi_j) = (\prod \psi_j) \circ (\prod \phi_j).$$

1.4 Relations

Formally, of course, a *relation* on a set X is just a subset R of $X \times X$. If $(\xi, \eta) \in R$ we write $\xi R \eta$ and say that ξ is R-related to η. Given ξ we denote by $R[\xi]$ the set of R-relatives of ξ, thus

$$R[\xi] = \{\eta \colon \xi R \eta\}.$$

Similarly, if H is a subset of X we denoted by $R[H]$ the set of R-relatives of points of H, so that $R[H]$ is the union of the subsets $R[\xi]$ for $\xi \in H$. Note that if $\{H_j\}$ if a family of subsets of X then

$$R[\bigcup H_j] = \bigcup R[H_j], \qquad R[\bigcap H_j] \subset \bigcap R[H_j].$$

The *identity relation* on the set X is just the diagonal ΔX of $X \times X$. A relation R which contains ΔX is said to be *reflexive*. If R is a relation on X the *reverse relation* R^{-1} is given by

$$\xi R^{-1} \eta \quad \text{if and only if} \quad \eta R \xi.$$

We say that R is *symmetric* if $R = R^{-1}$; for example, ΔX is symmetric.

The *composition* of relations R, S on the set X is the relation $R \circ S$, given by:

$$\xi (R \circ S) \eta \quad \text{if and only if} \quad \xi S \zeta \text{ and } \zeta R \eta \text{ for some } \zeta.$$

Composition of relations is associative and so bracketing is unnecessary for repeated compositions such as

$$R^n = R \circ \cdots \circ R \quad (n \text{ factors}).$$

Note that $(R \circ S)^{-1} = S^{-1} \circ R^{-1}$. Also note that if H is a subset of X then

$$(R \circ S)[H] = R[S[H]].$$

Composition of relations is not generally commutative. For example (taking $X = \{0, 1, 2\}$) if $R = \{(1, 2)\}$ and $S = \{(0, 1)\}$ then $R \circ S = \{(0, 2)\}$ while $S \circ R$ is empty.

A relation R is said to be *transitive* if $R \circ R$ is contained in R. An equivalence relation R is reflexive, symmetric and transitive; in that case the set $R[\xi]$ is just the equivalence class of ξ. The set of equivalence classes is denoted by X/R. Thus to each equivalence relation R on X there corresponds a surjection $\pi \colon X \to X/R$, where $\pi(\xi) = \{R[\xi]\}$. The reader may wish to determine which of the relations on the real line shown in Figure 1.2 is transitive.

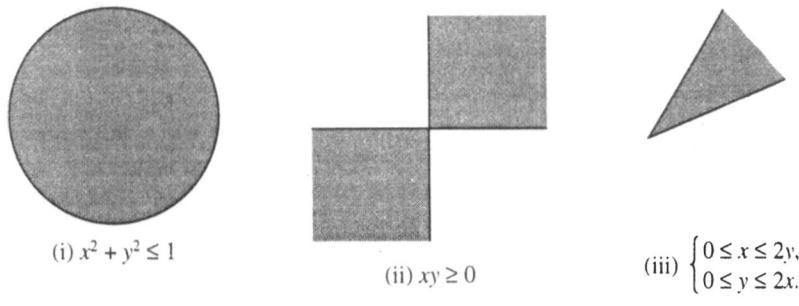

(i) $x^2 + y^2 \leq 1$

(ii) $xy \geq 0$

(iii) $\begin{cases} 0 \leq x \leq 2y, \\ 0 \leq y \leq 2x. \end{cases}$

Figure 1.2.

1.5 Axioms of topology

In most branches of mathematics the primary objects of study are sets with some kind of additional structure. In algebra, for example, the additional structure takes the form of one or more binary operations. In topology the additional structure consists of a family of subsets satisfying certain conditions:

Definition 1.1

A topology on a set X is a family of subsets of X, called open sets, such that:

   (i)   the empty set and the full set are open sets,
  (ii)   the intersection of a pair of open sets is an open set,
 (iii)   the union of any family of open sets is an open set.

Thus (ii) implies, by iteration, that the intersection of any finite family of open sets is an open set; in general, the intersection of an infinite family of open sets is not an open set.

By a *topological space* we mean a set X together with a topology $\mathcal{T}$ on X; usually X alone is sufficient notation. A *refinement* of the topology $\mathcal{T}$ is a topology $\mathcal{T}'$ on the same set X such that each open set of $\mathcal{T}$ is also an open set of $\mathcal{T}'$. In this situation we say that $\mathcal{T}'$ refines $\mathcal{T}$ or that $\mathcal{T}$ coarsens $\mathcal{T}'$. If the possibility that $\mathcal{T} = \mathcal{T}'$ is excluded we describe the refinement as *strict*. Let us have a few examples.

Definition 1.2

The discrete topology on the set X is the topology in which every subset is open.

In this situation we describe X as a *discrete space*. Clearly the discrete topology refines every other topology. Going to the other extreme we have

Definition 1.3

The trivial[1] topology on the set X is the topology in which the empty set and the full set are the only open sets.

In this situation we describe X as a *trivial space*. When X has fewer than two points the discrete topology and the trivial topology coincide; no other topology is possible. When X has two or more points, however, the discrete topology and the trivial topology are different, and there are other possible topologies as well.

For example, consider the case of the point-pair $\{0, 1\}$ (the choice of labels is immaterial). As well as the discrete topology and the trivial topology there are two others. In one of these $\{0\}$ is open but $\{1\}$ is not, in the other $\{1\}$ is open but $\{0\}$ is not. These are known as the *Sierpinski topologies*.

Thus there are four distinct topologies on a set with two points, and one can show that there are 29 distinct topologies on a set with three points, and so on. In the case of an infinite set the number of topologies is infinite. In Figure 1.3, which of the families of subsets of the three-point set constitute topologies, with the addition of the empty set, and which do not?

Here is another example which is different from the discrete topology in the case of an infinite set, and illustrates the point that the intersection of an infinite number of open sets is not necessarily an open set.

---

[1] The terms coarse topology and indiscrete topology are also in common use.

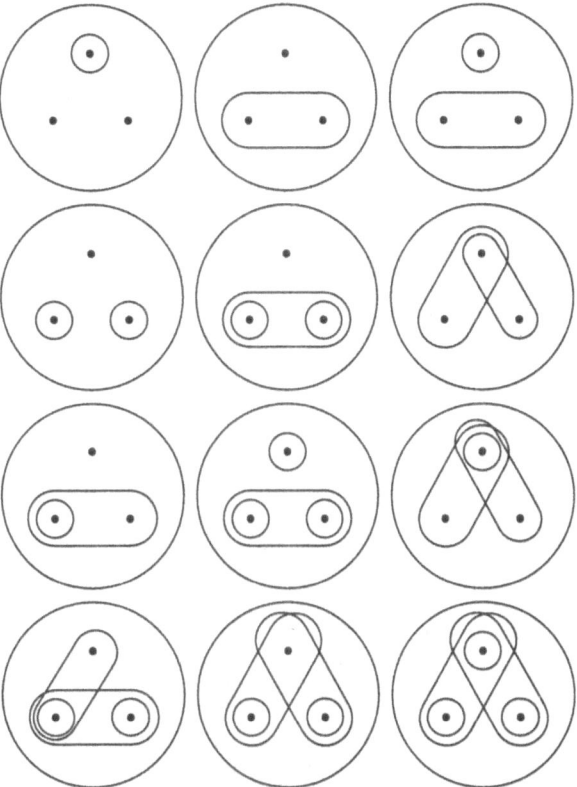

Figure 1.3.

Definition 1.4

The cofinite topology on the set X is the topology in which the open sets are the empty set and the cofinite sets, i.e. the complements of the finite sets.

In this situation we describe X as a *cofinite space*. Cocountable spaces may be defined similarly.

The examples so far given are valuable for illustrating points in the elementary theory since it is immediately clear as to whether or not a given subset is open, in a particular topology. Later we shall describe other, much more important, examples where the status of a given set may be less transparent.

With every family $\mathscr{F}$ of subsets of a set X there is associated a dual family $\mathscr{F}^*$, consisting of the complements of the members of $\mathscr{F}$. Any condition placed on the members of $\mathscr{F}$ can be transformed, according to the De Morgan laws, into a condition on the members of $\mathscr{F}^*$. In case $\mathscr{F}$ is a topology the members of the dual family are called *closed* sets, and we have

Definition 1.5

A topology on a set X is given by a family of subsets of X, called closed sets, such that:

  (i)   the empty set and the full set are closed sets,
 (ii)   the union of a pair of closed sets is a closed set,
(iii)   the intersection of any family of closed sets is a closed set.

It is generally agreed nowadays that a topology consists, as in Definition 1.1, of the family of open sets, rather than the dual family of closed sets; the latter are simply the complements of the open sets, and the conditions in Definition 1.5 describe how they behave. Thus (ii), by iteration, implies that the union of any finite family of closed sets is closed; in general, the union of an infinite family of closed sets is not a closed set.

In our first pair of examples, the discrete topology and the trivial topology, it happens that every open set is closed and every closed set is open. Consider, however, our third example, the cofinite topology, where the closed sets are the finite subsets and the full set. In the case of a finite set the cofinite topology is the same as the discrete topology, but in the case of an infinite set the topology cannot be discrete. In fact an infinite cofinite space contains sets which are closed and not open, sets which are open and not closed, sets which are neither open nor closed and, of course, sets which are both open and closed. Here the empty set and the full set are the only subsets which are both open and closed but in general one must expect to find others (see Chapter 9 for further details).

1.6 Closure and interior

We now turn to a series of results concerning subsets of topological spaces.

Definition 1.6

Let H be a subset of the topological space X. The closure Cl H of H is the intersection of the closed sets of X which contain H. The interior Int H of H is the union of the open sets of X which are contained in H.

Clearly Cl H is closed and Int H is open. Thus Cl $H = H$ if and only if H is closed, while Int $H = H$ if and only if H is open.

These remarks imply that closure and interior are idempotent operators, as in

Proposition 1.7

Let H be a subset of the topological space X. Then

$$Cl(Cl \ H) = Cl \ H, \qquad Int(Int \ H) = Int \ H.$$

Informally we may refer to Cl H as the smallest closed set of X containing H and to Int H as the largest open set of X contained in H. Note that

$$X - Cl \ H = Int(X - H);$$

i.e. the interior of the complement is the complement of the closure. Also note that if $H \subset K \subset X$ then Cl $H \subset$ Cl K and Int $H \subset$ Int K.

In the case of the trivial topology we have Cl $H = X$ for each non-empty H and Int $H = \emptyset$ for each non-full H, while in the case of the discrete topology we have Cl $H = H = $ Int H, for each subset H of X. In the case of the cofinite topology we have Cl $H = H$ when H is finite and Cl $H = X$ when H is infinite.

Definition 1.8

The subset H of the topological space X is dense in X if Cl $H = X$.

For example, in the case of the trivial topology every non-empty subset is dense, while in the case of the discrete topology no proper subset is dense. In the case of the cofinite topology on an infinite set the dense subsets are the infinite subsets.

So far we have said nothing at all about the status of the individual points (strictly speaking, one-point subsets) of a topological space. In general, these have no special status; like other subsets they may or may not be open sets, and may or may not be closed sets. However, the relation between the points and the topology is always of importance and this is the next subject for consideration.

Definition 1.9

Let H be a subset of the topological space X and let x be a point of X. If $x \in \operatorname{Int} H$ then x is an interior point of H. If $x \in \operatorname{Cl} H$ then x is an adherence[2] point of H.

Thus $\operatorname{Int} H$ consists of the interior points of H and $\operatorname{Cl} H$ consists of the adherence points of H. Our next result gives a useful criterion for a point to be an interior point or an adherence point.

In the statement of this result, and elsewhere, we will use the expression "open neighbourhood of" as a synonym for "open set containing". This is standard practice, although some topologists would write "neighbourhood" where we write "open neighbourhood". Thus an open neighbourhood of a point x is simply an open set containing x, so that an open set is an open neighbourhood of every point of itself. If the topology is discrete then any subset containing x, including $\{x\}$ itself, is an open neighbourhood of x. If the topology is trivial then the only open neighbourhood of x is the full set X.

Proposition 1.10

Let H be a subset of the topological space X and let x be a point of X. Then

(i)   *x is an interior point of H if and only if H contains some open neighbourhood of x,*
(ii)   *x is an adherence point of H if and only if H intersects every open neighbourhood of x.*

Note that (i) follows from (ii), and vice versa, by taking complements. It is therefore sufficient to give details for (ii).

Suppose that every open neighbourhood of x intersects H. The open set $X - \operatorname{Cl} H$ does not intersect H and so does not contain x, hence x is an adherence point of H.

Suppose that some open neighbourhood U of x does not intersect H. Then $X - U$ is closed and so contains $\operatorname{Cl} H$. But $x \in U$ and so x is not an adherence point of H.

---

[2] Various other terms, such as closure point, are used in the literature instead of adherence point.

Corollary 1.11

Let H be a subset of the topological space X. Then

(i) *H is an open set of X if and only if each point of H admits an open neigh-bourhood contained in H,*

(ii) *H is a closed set of X if and only if each point of X − H admits an open neighbourhood contained in X − H.*

1.7 Generating families

Suppose that we are given a topology on a set X and wish to refine it by making a particular subset H an additional open set. Unless X has the trivial topology it is not sufficient simply to add H to the family of open sets. If the axioms are to be satisfied we also have to add each of the sets $U \cup (V \cap H)$, where U and V run through the open sets of the original topology. Because the operations of union and intersection distribute with respect to each other it is easy to see that after enlarging the family of open sets in this way we again have a topology, the coarsest topology which contains H as well as the open sets of the original topology.

So now let X be a set and let Γ be a family of subsets of X. By the *topology generated* by Γ we mean the coarsest topology in which all the members of Γ are open sets. Thus as well as the empty set and the full set we have to take, first, all finite intersections of members of Γ and, secondly, all (unrestricted) unions of these finite intersections.

An important simplification occurs if the first step, that of taking finite inter-sections, can be omitted, so that it is only necessary to take unions of members of the family. A convenient sufficient condition for this is given in

Definition 1.12

Let Γ be a family of subsets of the set X. The family is complete if, whenever two members of Γ intersect, the intersection is also a member of Γ.

For example, the one-point subsets of X satisfy this condition and generate the discrete topology.

Of course the first step can also be omitted if, in Definition 1.12, the inter-section is a union of members of Γ. In that case Γ is said to be a *base* for the topology it generates.

The real line $\mathbb{R}$ provides a good illustration of the use of generating families. In fact the rich variety of structures enjoyed by the real line enables it to be topologized by various methods, some of which lead to the same final result. The first method we shall use depends on the order relation.

1.8 Ordered sets

Recall that an *order* on a set X is a transitive relation $<$ on X such that for each pair of points ξ, η of X there holds one and only one of the following:

$$\text{(i) } \xi < \eta, \qquad \text{(ii) } \xi = \eta, \qquad \text{(iii) } \eta < \xi.$$

As usual we write $\eta > \xi$ when $\xi < \eta$. We also write $\xi \leq \eta$ when $\xi < \eta$ or $\xi = \eta$, and $\xi \geq \eta$ when $\xi > \eta$ or $\xi = \eta$. An *ordered set*[3] is a set X together with an order $<$; usually X alone is sufficient notation.

Let X be an ordered set. A subset H of X is said to be an *interval* if the following condition is satisfied. Suppose that $\alpha < \xi < \beta$, where α, β, $\xi \in X$; if α, $\beta \in H$ then $\xi \in H$. Here are some examples.

First, the full set X and the empty set $\varnothing$ are intervals. Second, for each point $\alpha \in X$ the intervals

$$(-\infty, \alpha) = \{\xi \colon \xi < \alpha\}, \qquad (\alpha, +\infty) = \{\xi \colon \xi > \alpha\}$$

are defined. These are called *open* intervals, to distinguish them from

$$(-\infty, \alpha] = \{\xi \colon \xi \leq \alpha\}, \qquad [\alpha, +\infty) = \{\xi \colon \xi \geq \alpha\},$$

which are called *closed* intervals. The term *ray* is sometimes used to describe open or closed intervals of this type.

Third, for each pair of points α, $\beta \in X$ the open intervals

$$(\alpha, \beta) = \{\xi \colon \alpha < \xi < \beta\},$$

and the closed intervals

$$[\alpha, \beta] = \{\xi \colon \alpha \leq \xi \leq \beta\},$$

are defined, also the *half-open* (or *half-closed*) intervals

$$[\alpha, \beta) = \{\xi \colon \alpha \leq \xi < \beta\},$$

$$(\alpha, \beta] = \{\xi \colon \alpha < \xi \leq \beta\}.$$

---

[3] The terms linearly ordered set and partially ordered set are also in common use.

When $\alpha = \beta$ the open and half-open types of interval are empty, while the closed interval reduces to $\{\alpha\}$. When $\alpha > \beta$ all four types of interval are empty.

Note that the four types of interval just specified can be obtained as intersections of rays. Thus

$$(\alpha, \beta) = (-\infty, \beta) \cap (\alpha, +\infty),$$

$$[\alpha, \beta] = (-\infty, \beta] \cap [\alpha, +\infty),$$

and so on.

Consider the family of open rays $(-\infty, \alpha)$, $(\alpha, +\infty)$, where $\alpha \in X$. The family satisfies the completeness condition and generates a topology on X, called the *interval topology*. Note that the open intervals are indeed open, in this topology. This is clear for (α, β) itself, while for the open rays it follows from the observation that $(\alpha, +\infty)$ is the union of the generating sets (α, β) where β runs through all points of X, and similarly in the case of $(-\infty, \alpha)$. Hence the closed intervals $[\alpha, \beta]$ are indeed closed, since the complement of $[\alpha, \beta]$ is the union of the open intervals $(-\infty, \alpha)$ and $(\beta, +\infty)$. In particular, taking $\alpha = \beta$, we see that the individual points of X are closed, in the interval topology.

Of course the order on the set X can be used to produce a topology on X in other ways. In discussing these however, we must be cautious about using the terms open interval, closed interval and so forth, since they refer to the situation in respect of the interval topology itself; we shall therefore place the terms open and closed in inverted commas.

The family of "closed" intervals of the form $[\alpha, \beta]$ contains the one-point sets and so simply generates the discrete topology. More interestingly each of the families of "half-open" intervals generates a topology, coarser than the interval topology, which is different from any we have previously considered. The topology generated by the family of intervals of the form $[\alpha, \beta)$ is called the *upper Sorgenfrey* topology, while the topology generated by the family of intervals of the form $(\alpha, \beta]$ is called the *lower Sorgenfrey* topology. These Sorgenfrey topologies[4] will be used to provide examples and counterexamples, from time to time, but they do not play an essential role in what is to follow.

Returning to the interval topology itself, in the special case of $\mathbb{R}$, let us examine the status of some subsets of $\mathbb{R}$ such as the subset $\mathbb{Q}$ of rationals and the subset $\mathbb{R} - \mathbb{Q}$ of irrationals. If ξ is rational then any open interval (α, β), where $\alpha < \xi < \beta$, contains irrational numbers and so is not contained in $\mathbb{Q}$;

---

[4] Also known as the upper and lower limit topologies.

therefore, $\mathbb{Q}$ is not open and $\mathbb{R} - \mathbb{Q}$ is not closed. Similarly, if ξ is irrational then any open interval (α, β), where $\alpha < \xi < \beta$, contains rational numbers and so is not contained in $\mathbb{R} - \mathbb{Q}$; therefore $\mathbb{R} - \mathbb{Q}$ is not open and $\mathbb{Q}$ is not closed. Observe, however, that for any real number ξ an open interval (α, β), where $\alpha < \xi < \beta$, contains both rational and irrational numbers, in other words (α, β) intersects both $\mathbb{Q}$ and $\mathbb{R} - \mathbb{Q}$. We conclude, therefore, that $\mathbb{Q}$ and $\mathbb{R} - \mathbb{Q}$ are dense subsets of $\mathbb{R}$.

1.9 Filters

A sequence in a set X is just a function assigning to each natural number n a point x_n of X, the nth term of the sequence. Sequences play a useful role in elementary real analysis, for example. However, for our purposes the simple notion of sequence turns out to be inadequate. There are two ways to deal with this problem.

One is to introduce generalized sequences or nets, in which the natural numbers are replaced by ordered sets. The other is to introduce the notion of filter. This is a rather more sophisticated replacement for the simple notion of sequence but it is considerably more versatile. Consequently we shall be using filters rather than nets and so a summary of the relevant theory is included for the convenience of the reader.

Definition 1.13

A filter $\mathscr{F}$ on a set X is a non-empty family of non-empty subsets of X such that:

(i)   each superset of a member of $\mathscr{F}$ is also a member of $\mathscr{F}$,
(ii) the intersection of each pair of members of $\mathscr{F}$ is also a member of $\mathscr{F}$.

The second condition implies, of course, that the intersection of a finite family of members of $\mathscr{F}$ is again a member of $\mathscr{F}$ and in particular is non-empty. However, this does not necessarily apply to infinite families. For example, the cofinite subsets of an infinite set form a filter $\mathscr{F}_0$ such that the intersection of all the members of $\mathscr{F}_0$ is empty.

The non-empty subsets of a set X generate filters in an obvious way. For example, all the subsets of X which contain a given point x of X form a filter ε_x; this is called the *principal filter* generated by x. Again all the subsets of X which contain a given non-empty subset H of X form a filter. These examples have the property that the intersection of all the members of the filter is itself a member of the filter.

For another type of example we turn to the notion of (infinite) sequence. A sequence of points of a set X is a function $\alpha: \mathbb{N} \to X$, where $\mathbb{N}$ denotes the ordered set of natural numbers. It is usually convenient to write the sequence in the form $\{x_n: n = 1, 2, \ldots\}$, or simply $\langle x_n \rangle$, where $x_n = \alpha(n)$ is the nth term of the sequence. Subsequences are defined by precomposing α with an order-preserving injection $\sigma: \mathbb{N} \to \mathbb{N}$.

Definition 1.14

Let $\langle x_n \rangle$ be a sequence of points of the set X. The elementary filter associated with $\langle x_n \rangle$ is the filter consisting of subsets M of X for which there exists an integer k such that $x_n \in M$ whenever $n \geq k$.

Note that it is necessary to know more than just the set $\{x_n\}$ in order to define the filter. Also note that if $\langle x_n \rangle$ and $\langle x_n' \rangle$ are sequences such that for some integer k we have $x_n = x_n'$ whenever $n \geq k$ then the elementary filter associated with $\langle x_n \rangle$ coincides with the elementary filter associated with $\langle x_n' \rangle$.

For example, take $X = \mathbb{N}$ and consider the sequence $\langle n \rangle$ of which the nth term is n. The elementary filter associated with this sequence consists of the cofinite subsets of the set $\mathbb{N}$.

Definition 1.15

A non-empty family of non-empty subsets of the set X is a filter base if the intersection of every pair of members of the family contains a member of the family.

If this condition is satisfied then the family generates a filter on X, by taking supersets of members of the base. Note that two filter bases on the same set generate the same filter if, and only if, each member of the first base contains a member of the second, and vice versa.

Of course every filter is generated by itself, regarded as a base. Also the principal filter ε_x, where x is a point of X, is generated by $\{x\}$.

Let $\mathcal{F}$, $\mathcal{G}$ be filters on X and suppose that every member of $\mathcal{F}$ intersects every member of $\mathcal{G}$. Then the family of intersections $M \cap N$, where $M \in \mathcal{F}$ and $N \in \mathcal{G}$, has the property that the intersection of a pair of members of the family is again a member of the family. In particular, the family constitutes a filter base and the filter thus generated is called the *filter generated by* $\mathcal{F}$ and $\mathcal{G}$. It cannot be defined if there exist disjoint members of $\mathcal{F}$ and $\mathcal{G}$.

Let $\mathscr{F}$ be a filter on the set X. For each set X' and function $\phi\colon X \to X'$ a filter $\phi_*\mathscr{F}$ on X' is defined by taking the direct images of the members of $\mathscr{F}$ as a filter base. In case X' is a superset of X and ϕ the inclusion we refer to $\phi_*\mathscr{F}$ as the extension of $\mathscr{F}$ to X'.

In the other direction, let X' be a set and $\phi\colon X' \to X$ a function. If the inverse images of the members of $\mathscr{F}$ are all non-empty, as is always the case when ϕ is surjective, then the conditions for a filter base are satisfied and the filter on X' thereby generated is called the *induced* filter and denoted by $\phi^*\mathscr{F}$. In case X' is a subset of X and ϕ the inclusion we refer to $\phi^*\mathscr{F}$ as the trace of $\mathscr{F}$ on X', emphasizing that it is only defined when each member of $\mathscr{F}$ has non-empty intersection with X'.

Definition 1.16

Let $\mathscr{F}$ be a filter on the set X. A refinement of $\mathscr{F}$ is a filter $\mathscr{F}'$ on X such that each member of $\mathscr{F}$ is also a member of $\mathscr{F}'$.

In this situation we say that $\mathscr{F}'$ refines $\mathscr{F}$, or that $\mathscr{F}$ is refined by $\mathscr{F}'$. When the possibility that $\mathscr{F} = \mathscr{F}'$ is to be excluded we describe the refinement as *strict*.

For example, let $\mathscr{F}$ be the elementary filter associated with a sequence $\langle x_n \rangle$ in X. The elementary filter associated with any subsequence of $\langle x_n \rangle$ is a refinement of $\mathscr{F}$. Also if $\mathscr{F}'$ is an elementary filter associated with a sequence $\langle x'_n \rangle$ in X, and if $\mathscr{F}'$ refines $\mathscr{F}$, then $\mathscr{F}'$ is the elementary filter associated with a subsequence of $\langle x_n \rangle$.

For another example, let $\mathscr{F}$ and $\mathscr{G}$ be filters on X such that each member of $\mathscr{F}$ intersects each member of $\mathscr{G}$. Then the filter generated by the intersections is a refinement of both $\mathscr{F}$ and $\mathscr{G}$.

For a third example, let $\phi\colon X \to Y$ be a function, where X and Y are sets. Then $\phi^*\phi_*\mathscr{F}$ is refined by $\mathscr{F}$, for each filter $\mathscr{F}$ on X. And $\phi_*\phi^*\mathscr{G}$ refines $\mathscr{G}$ for each filter $\mathscr{G}$ on Y such that $\phi^*\mathscr{G}$ is defined.

Refinement imposes a partial order on the collection of filters on the set X. A maximal member of the collection is called an *ultrafilter*. Using Zorn's lemma one can show that every filter admits an ultrafilter refinement (in general, far from unique). Principal filters are obviously ultrafilters but ultrafilters in general are somewhat mysterious objects.

Proposition 1.17

Let $\mathscr{F}$ be a filter on the set X. Then $\mathscr{F}$ is an ultrafilter if and only if whenever $M \cup N \in \mathscr{F}$, where M, N are subjects of X, either $M \in \mathscr{F}$ or $N \in \mathscr{F}$.

For suppose that $M \cup N \in \mathscr{F}$ but $M \notin \mathscr{F}$ and $N \notin \mathscr{F}$. Let $\mathscr{F}'$ be the family of subsets N' of X such that $M \cup N' \in \mathscr{F}$. Then $\mathscr{F}'$ is a refinement of $\mathscr{F}$ and since $N \notin \mathscr{F}$ the refinement is strict, hence $\mathscr{F}$ is not an ultrafilter.

Conversely, let $\mathscr{F}'$ be a strict refinement of $\mathscr{F}$. If M is a member of $\mathscr{F}'$ but not of $\mathscr{F}$ then $M \cup (X - M) = X \in \mathscr{F}$, and so $X - M \in \mathscr{F}$, by the condition. But this implies $X - M \in \mathscr{F}'$, since $\mathscr{F}'$ is a refinement of $\mathscr{F}$, and then $M \cap (X - M) = \varnothing$, contrary to the definition of filter.

Hence by induction we obtain

Proposition 1.18

Let $M_1, \ldots, M_n$ be subsets of the set X, and let $\mathscr{F}$ be an ultrafilter on X. If the union of $M_1, \ldots, M_n$ is a member of $\mathscr{F}$ then one of $M_1, \ldots, M_n$ is a member of $\mathscr{F}$.

In the case of a finite set every ultrafilter is necessarily principal. In the case of an infinite set a special role is played by the filter $\mathscr{F}_0$ consisting of cofinite subsets mentioned earlier. Since $\mathscr{F}_0$ is obviously not principal neither is any ultrafilter refinement of $\mathscr{F}_0$. In fact, any non-principal ultrafilter $\mathscr{F}$ must be a refinement of $\mathscr{F}_0$. For if $M \in \mathscr{F}_0$ then either $M \in \mathscr{F}$ or the finite set $X - M \in \mathscr{F}$. But if $X - M \in \mathscr{F}$ then $\{x\} \in \mathscr{F}$ for some $x \in X - M$, by Proposition 1.18, which is a contradiction, since $\mathscr{F}$ is non-principal.

1.10 Neighbourhoods

So far we have been using the term "open neighbourhood of" as a synonym for "open set containing". From now on I also wish to use the term "neighbourhood" as a synonym for "superset of an open neighbourhood". Thus if x is a point of the topological space X a subset N of X is a *neighbourhood* of x if x belongs to the interior of N, and similarly if x is replaced by a subset of X.

If N itself is open the term "open neighbourhood" has not changed its meaning. If N is closed we use the term "*closed neighbourhood*".

For each point x of the topological space X the family of neighbourhoods of x forms a filter $\mathscr{N}_x$, the neighbourhood filter of x. Since each neighbourhood contains x itself, the principal filter ε_x is a refinement of $\mathscr{N}_x$. A filter base for $\mathscr{N}_x$ is called a neighbourhood base at x. Thus a family $\mathscr{B}_x$ of neighbourhoods of x is a neighbourhood base if every neighbourhood of x contains a basic neighbourhood, i.e. a member of $\mathscr{B}_x$. For example, the open neighbourhoods form a neighbourhood base although the closed neighbourhoods in general do not. We shall discuss this point further in Chapter 6.

By way of illustration, consider the real line $\mathbb{R}$ with the interval topology. Since the open intervals (α, β) generate the topology, a neighbourhood base at a given point ξ consists of those open intervals (α, β) such that $\alpha < \xi < \beta$. However, this is not the only neighbourhood base in this topology. Another, which is generally more convenient in practice, consists of the open intervals $(\xi - \varepsilon, \xi + \varepsilon)$ for all $\varepsilon > 0$. Or, more generally, we can use open intervals of the form $(\xi - \lambda\varepsilon, \xi + \mu\varepsilon)$, for fixed positive λ, μ and all $\varepsilon > 0$. Yet another possibility is to use open intervals of the same form but with ε restricted to the positive rationals. Each of these constitutes a neighbourhood base in the interval topology.

Returning to the general case we now observe that Proposition 1.10 remains true for any neighbourhood base at the point in question, rather than the open neighbourhoods specifically. Since this is the form in which the result is most commonly used we restate it as

Proposition 1.19

Let H be a subset of the topological space X and let x be a point of X. In terms of a given neighbourhood base at x then

(i)   *x is an interior point of H if and only if H contains some basic neighbourhood of x,*
(ii) *x is an adherence point of H if and only if H intersects every basic neighbourhood of x.*

The corollary can also be reformulated in a similar fashion.

Neighbourhood filters at different points are not unrelated since, after all, the same open set is a neighbourhood of every one of its own points. In fact, the collection of filters $\mathcal{N}_x$, for all $x \in X$, satisfies the following condition.

Definition 1.20

Let X be a set and let $\{\mathcal{N}_x : x \in X\}$ be a collection of filters on X, such that ε_x is a refinement of $\mathcal{N}_x$ for each x. The collection is coherent if for each point x of X and each member N of $\mathcal{N}_x$ there exists a member N' of $\mathcal{N}_x$ such that N is a member of $\mathcal{N}_{x'}$, for each point x' of N'.

Proposition 1.21

Let X be a set and let $\{\mathcal{N}_x : x \in X\}$ be a coherent collection of filters on X, as in Definition 1.20. Let $\mathcal{T}$ be the family of subsets U of X which satisfy the *condition: $U \in \mathcal{N}_x$ whenever $x \in U$. Then $\mathcal{T}$ constitutes a topology on X such that $\mathcal{N}_x$, for each point x, is the neighbourhood filter of x defined by the topology.*

Let us refer to the members U of $\mathcal{T}$, without prejudice, as open sets. We first have to verify the axioms of topology. Since the empty set and the full set are obviously open, we only have to check the axioms for intersection and union.

Let U, V be open. If $x \in U \cap V$ then U, $V \in \mathcal{N}_x$, since $x \in U$ and $x \in V$, hence $U \cap V \in \mathcal{N}_x$, and so $U \cap V$ is open.

Let $\{U_j\}$ be a family of open sets. If $x \in \bigcup U_j$ then $x \in U_j$ for some j and so $U_j \in \mathcal{N}_x$. But $\bigcup U_j$ is a superset of U_j and so $\bigcup U_j \in \mathcal{N}_x$. Therefore $\bigcup U_j$ is open, completing the verification.

It remains to be shown that the neighbourhood system determined by the topology is the same as the coherent collection which determined the topology. If N is a neighbourhood of the point x then $x \in U \subset N$ for some open U. Now $x \in U$ implies $U \in \mathcal{N}_x$ and so $N \in \mathcal{N}_x$, since N is a superset of U. Thus each neighbourhood of x is a member of $\mathcal{N}_x$; conversely, each member N of $\mathcal{N}_x$ is a neighbourhood of x. Define U to be the set of all points $y \in X$ such that $N \in \mathcal{N}_y$. Clearly x is one of these points, and so $x \in U$. Also if $N \in \mathcal{N}_y$ then $y \in N$, and so $U \subset N$. I assert that U is open. For if $y \in U$ then $N \in \mathcal{N}_y$ and so, by Definition 1.20, there exists a member N' of $\mathcal{N}_y$ such that $N \in \mathcal{N}_{y'}$ whenever $y' \in N'$. But $N \in \mathcal{N}_{y'}$ implies $y' \in U$, by definition of U, so $N' \subset U$ and so $U \in \mathcal{N}_y$ as required. This completes the proof.

1.11 Metric spaces

A coherent collection of filters, as in Definition 1.20, is generally known as a *fundamental system of neighbourhoods*. This approach to topology is particularly well suited to the case when the set in question is equipped with a metric. In case the reader is not already familiar with the notion of metric space, we recall the essential features, beginning with

Definition 1.22

A metric on a set X is a non-negative real-valued function $\rho\colon X \times X \to \mathbb{R}$ such that

(i) $\rho(\xi, \xi) = 0$ for all $\xi \in X$,

(ii) $\rho(\xi, \eta) = \rho(\eta, \xi)$ for all $\xi, \eta \in X$,

(iii) $\rho(\xi, \zeta) \le \rho(\xi, \eta) + \rho(\eta, \zeta)$ for all $\xi, \eta, \zeta \in X$ (triangle inequality),

(iv)   if $\rho(\xi, \eta) = 0$, for some $\xi, \eta \in X$, then $\xi = \eta$.

A set X together with a metric ρ is called a *metric space*. Usually X alone is sufficient notation.

For example, the *discrete metric* on X is given by $\rho(\xi, \eta) = 0$ if $\xi = \eta$, $\rho(\xi, \eta) = 1$ otherwise.

For another example, the *euclidean metric* on the real line $\mathbb{R}$ is given by

$$\rho(\xi, \eta) = |\xi - \eta|.$$

More generally, the euclidean metric on the real n-space $\mathbb{R}^n$ is given by

$$\rho((\xi_1, \ldots, \xi_n), (\eta_1, \ldots, \eta_n)) = ((\xi_1 - \eta_1)^2 + \cdots + (\xi_n - \eta_n)^2)^{1/2}.$$

Other metrics on $\mathbb{R}^n$ (we take $n = 2$ for simplicity) include

(i) $\rho((x_1, x_2), (y_1, y_2)) = |x_1 - y_1| + |x_2 - y_2|,$
(ii) $\rho((x_1, x_2), (y_1, y_2)) = \max\{|x_1 - y_1|, |x_2 - y_2|\}.$

To get some feeling for these different metrics on $\mathbb{R}^2$ it may be helpful to draw the "unit circle" with respect to each of them, i.e. the set of points $x \in \mathbb{R}^2$ such that $\rho(x, 0) = 1$. For the euclidean metric the unit circle is the usual one. For the discrete metric it is the punctured plane $\mathbb{R}^2 - \{0\}$. For the last two metrics listed above, the unit circle is the lozenge-shaped figure on the left of Figure 1.4 in the case of (i) and the square on the right in the case of (ii).

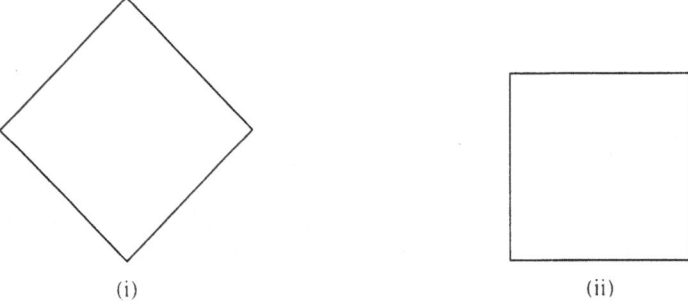

(i)   (ii)

Figure 1.4.

If the examples we have so far mentioned were the only examples of metric spaces the motivation for studying the general concept might not seem very strong. But in fact there are a multitude of examples of different types which arise very naturally in functional analysis. Here are just a few.

Consider the set $\mathbb{H}$ of sequences $\langle x_n \rangle$ of real numbers such that the series $\sum x_n^2$ converges. A metric ρ on $\mathbb{H}$ is defined by

$$\rho(\langle x_n \rangle, \langle y_n \rangle) = \left\{ \sum_{n=1}^{\infty} (x_n - y_n)^2 \right\}^{1/2},$$

and $\mathbb{H}$ with this metric is called the *Hilbert (sequence) space.*

For any non-empty set X let $B(X)$ denote the set of bounded real-valued functions $X \to \mathbb{R}$. The *supremum metric* ρ on $B(X)$ is given by

$$\rho(f, g) = \sup_{\xi \in X} |f(\xi) - g(\xi)|,$$

where $f, g \colon X \to \mathbb{R}$ are bounded.

For any non-empty topological space X let $C(X)$ denote the set of continuous real-valued functions $X \to \mathbb{R}$ and let $C^*(X) = B(X) \cap C(X)$ denote the subset of bounded functions. We can give $C^*(X)$ the supremum metric, as above, but this is by no means the only possibility.

For example, take $X = I = [0, 1]$. One alternative to the supremum metric is the metric ρ given by

$$\rho(f, g) = \int_0^1 |f(\xi) - g(\xi)| \, d\xi$$

and another is the metric ρ given by

$$\rho(f, g) = \left\{ \int_0^1 (f(\xi) - g(\xi))^2 \, d\xi \right\}^{1/2}.$$

Returning to the general case, let X be a metric space with metric ρ. We introduce the notation

$$B_\varepsilon(x) = \{\xi \colon \rho(\xi, x) \leq \varepsilon\},$$

where $\varepsilon \geq 0$, for the *closed ε-ball* at x and

$$U_\varepsilon(x) = \{\xi \colon \rho(\xi, x) < \varepsilon\},$$

where $\varepsilon > 0$, for the *open ε-ball*. The justification for the use of the terms open and closed in this context will emerge in a moment. Note that in the case of the real line $\mathbb{R}$, with the euclidean metric, the closed ε-ball is just the closed interval $[x - \varepsilon, x + \varepsilon]$ and the open ε-ball is just the open interval $(x - \varepsilon, x + \varepsilon)$.

Clearly, the open ε-balls at a given point x of X, for all positive ε, generate a filter $\mathcal{N}_x$ on the metric space X. I assert that the collection of filters $\mathcal{N}_x$, for all $x \in X$, satisfies the coherence condition (Definition 1.20). Since it is sufficient to show that the condition is satisfied for members of the filter base, consider the open ε-ball $U_\varepsilon(x)$, where $\varepsilon > 0$ and $x \in X$. If $\xi \in U_\varepsilon(x)$ then $\rho(\xi, x) < \varepsilon$ and so $\delta = \varepsilon - \rho(\xi, x) > 0$. Therefore $U_\delta(\xi) \subset U_\varepsilon(x)$, by the triangle inequality. Since $U_\delta(\xi)$ is a member of $\mathcal{N}_\xi$ this establishes coherence. We conclude, therefore, that the collection of filters determines a topology on X in which $\mathcal{N}_x$ is the neighbourhood filter of x, for each point x. We refer to this as the *metric topology.*

For example, in the case of the discrete metric $\mathcal{N}_x$ reduces to the principal filter ε_x and the metric topology is just the discrete topology.

For another example, consider the real line $\mathbb{R}$, with the euclidean metric. At a given point x of $\mathbb{R}$ each basic neighbourhood of the metric topology is also a basic neighbourhood of the interval topology, while each basic neighbourhood (α, β) of the interval topology contains the basic neighbourhood $(x - \varepsilon, x + \varepsilon)$ of the metric topology, where $\varepsilon = \min(\beta - x, x - \alpha)$. Therefore the neighbourhood filters are the same, in both cases, and so the interval topology and the euclidean topology coincide. Since the euclidean topology is available in higher dimensions while the interval topology is not we shall tend to use the term euclidean topology from now on, even in the case of the real line.

Returning to the general case it should be noted that although the metric topology depends on the choice of metric it is nevertheless true that different metrics may determine the same topology. For example, if ρ is a metric on the set X then 2ρ is also a metric on X; the open ε-balls in the case of ρ are the same as the open 2ε-balls in the case of 2ρ, hence the metric topologies are the same since the neighbourhood filters are the same.

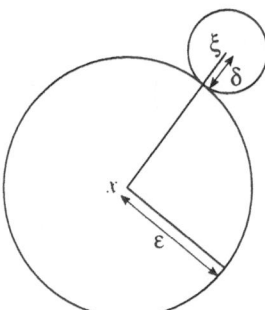

Figure 1.5.

In the metric topology the open ε-balls ($\varepsilon > 0$) are obviously open sets. To see that the closed ε-balls ($\varepsilon \geq 0$) are closed sets we use the test in Corollary 1.11: if $\xi \notin B_\varepsilon(x)$ then $\rho(\xi, x) > \varepsilon$ and so $U_\delta(\xi) \subset X - B_\varepsilon(x)$, where $\delta = \rho(\xi, x) - \varepsilon$, as illustrated in Figure 1.5. In particular (taking $\varepsilon = 0$) we see that points are closed, in the metric topology. This implies, of course, that any topology in which points are not closed – for example, the Sierpinski topology on the point-pair – cannot be obtained from a metric.

According to Proposition 1.19, the necessary and sufficient condition for a point x to adhere to a subset H of a metric space X is that $U_\varepsilon(x)$ intersects H for all $\varepsilon > 0$. We can restate this as follows. Assuming H is non-empty the set

$$\{\rho(x, \xi) : \xi \in H\}$$

of real numbers is bounded below (by zero) and so the infimum is defined. We denote the infimum by $\rho(x, H)$ and note that $\rho(x, H) = 0$ if and only if x adheres to H. If, as is natural, we think of $\rho(x, H)$ as the distance from x to H then the closure of H is just the set of points whose distance from H is zero.

It should be noted that the closure of the open ball $U_\varepsilon(x)$ is generally a proper subset of the closed ball $B_\varepsilon(x)$, although $\operatorname{Cl} U_\varepsilon(x) = B_\varepsilon(x)$ in the case of $\mathbb{R}^n$. For example, if X has the discrete metric then $\operatorname{Cl} U_1(x) = \operatorname{Cl} \{x\} = \{x\}$ while $B_1(x) = X$. However the closed balls at x always form a neighbourhood base at x.

It is usually not difficult, in practice, to decide the status of a given subset of a metric space. For example, consider the real n-space R^n, with the euclidean metric, and write $\|(x_1, \ldots, x_n)\| = (x_1^2 + \cdots + x_n^2)^{1/2}$ in the usual way. Among the subsets of $\mathbb{R}^n$ there are several which will often be featured in our work as it progresses. Using the criteria of Corollary 1.11 it is not difficult to decide the status of each of them. Notable examples of closed sets include the closed n-ball

$$B^n = \{x \in \mathbb{R}^n : \|x\| \le 1\}$$

and the $(n-1)$-sphere

$$S^{n-1} = \{x \in \mathbb{R}^n : \|x\| = 1\}.$$

Notable examples of open sets include the open n-ball

$$U^n = B^n - S^{n-1} = \{x \in \mathbb{R}^n : \|x\| < 1\}$$

and the punctured open n-ball

$$U^n - \{0\} = \{x \in \mathbb{R}^n : 0 < \|x\| < 1\}.$$

Further examples will be given in the exercises.

1.12 Filters on topological spaces

Recall that in a topological space X, a point x of X is said to adhere to a subset H of X if x belongs to the closure $\operatorname{Cl} H$, i.e. if every neighbourhood of x intersects H. We extend this to filters as follows.

Definition 1.23

Let $\mathscr{F}$ be a filter on the topological space X. The point x of X adheres to $\mathscr{F}$ if x is an adherence point of every member of $\mathscr{F}$.

The condition implies, incidentally, that $\mathscr{F}$ and $\mathscr{N}_x$ generate a filter which refines them both. Note that if x adheres to $\mathscr{F}$ then x also adheres to every filter which is refined by $\mathscr{F}$. Note also that if $\mathscr{F}$ is generated by a filter base then x adheres to $\mathscr{F}$ if x adheres to every member of the filter base. In particular, if $\mathscr{F}$ is the principal filter generated by a non-empty subset H of X then the adherence set of $\mathscr{F}$ is just the closure Cl H of H.

Obviously the set of adherence points of a filter $\mathscr{F}$ is just the intersection of the closed members of $\mathscr{F}$. Suppose that X is an infinite set with cofinite topology. Then the adherence set of every filter $\mathscr{F}$ on X is non-empty. If every member of $\mathscr{F}$ is infinite then the adherence set is the full set.

The notion of convergence, and the associated notion of limit, is familiar from the theory of sequences of real numbers. It is easy enough to generalize this to sequences in any topological space, as in

Definition 1.24

A sequence $\langle x_n \rangle$ of points of the topological space X converges to a point x of X if for each neighbourhood U of x there exists an integer k such that $x_n \in U$ whenever $n \geq k$.

It is sufficient, of course, if the condition is satisfied for each member U of a neighbourhood base at x. For example, if X has the discrete topology then $\langle x_n \rangle$ converges to x if and only if there exists an integer k such that $x_n = x$ whenever $n \geq k$. More generally, if X is a metric space then $\langle x_n \rangle$ converges to x, in the metric topology, if and only if for each positive ε there exists an integer k such that $x_n \in U_\varepsilon(x)$ whenever $n \geq k$.

If $\langle x_n \rangle$ converges to x, as in Definition 1.24, we describe x as a *limit* of $\langle x_n \rangle$ and write $x_n \to x$. Limits, when they exist, are not necessarily unique. For example, if X has the trivial topology then every sequence in X converges to every point of X. However, limits are clearly unique in the metric case.

In general, sequences do not carry enough information about the topology for it to be possible to develop a satisfactory theory of convergence. The difficulty is overcome if we replace convergent sequences by convergent filters, as follows.

Definition 1.25

Let $\mathscr{F}$ be a filter on the topological space X. The point x of X is a limit point of $\mathscr{F}$ if $\mathscr{F}$ is a refinement of the neighbourhood filter $\mathscr{N}_x$ of x.

When this condition is satisfied we say that $\mathscr{F}$ converges to x, and write $\mathscr{F} \to x$. For example, $\mathscr{N}_x \to x$ for each $x \in X$. Note that if $\mathscr{F}$ converges to x then so does any refinement of $\mathscr{F}$. In other words limit points are inherited by refinements, whereas for adherent points it goes the other way.

The relation between convergence of filters and convergence of sequences is explained in

Proposition 1.26

A sequence $\langle x_n \rangle$ of points of the topological space X converges to x, in the sense of Definition 1.24, if and only if the elementary filter associated with $\langle x_n \rangle$ converges to x, in the sense of Definition 1.25.

The proof is an immediate consequence of the definition of the term elementary filter (Definition 1.14).

Limit points when they exist, are not in general unique. For example, if X is a trivial space every filter on X converges to every point of X.

Our next result elucidates the relationship between adherence points and limit points.

Proposition 1.27

Let $\mathscr{F}$ be a filter on the topological space X and let x be a point of X. Then x is an adherence point of $\mathscr{F}$ if and only if x is a limit point of some refinement of $\mathscr{F}$.

For if x adheres to $\mathscr{F}$ then each member of the neighbourhood filter $\mathscr{N}_x$ of x intersects every member of $\mathscr{F}$; hence the common refinement $\mathscr{G}$ of $\mathscr{N}_x$ and $\mathscr{F}$ is defined and has x as a limit point. Conversely, if x is a limit point of some refinement $\mathscr{G}$ of $\mathscr{F}$ then each neighbourhood of x is a member of $\mathscr{G}$ and so intersects every member of $\mathscr{F}$; thus x is an adherence point of $\mathscr{F}$. In case $\mathscr{F}$ is an ultrafilter strict refinement of $\mathscr{F}$ cannot occur and so we deduce

Corollary 1.28

For an ultrafilter $\mathscr{F}$ on the topological space X each adherence point of $\mathscr{F}$ is also a limit point of $\mathscr{F}$.

Exercises

1.1. Let $K \subset H \subset X$, where X is a topological space. Show that $H - K$ is open if H is open and K is closed, while $H - K$ is closed if H is closed and K is open.

1.2. Show that if U is an open set of the topological space X then
$$U \cap \mathrm{Cl}\, H \subset \mathrm{Cl}(U \cap H)$$
for each subset H of X.

1.3   Show that for any open set U of the topological space X
$$\mathrm{Cl}\, U = \mathrm{Cl}(\mathrm{Int}(\mathrm{Cl}\, U)).$$

1.4. A subset of a topological space is said to be *regularly open* if it coincides with the interior of its closure, *regularly closed* if it coincides with the closure of its interior. Show that the complement of a regularly open set is regularly closed, and vice versa. Also show that the interior of the closure of a subset is always regularly open, and that the closure of the interior of a subset is always regularly closed.

1.5. In the real line $\mathbb{R}$, with the euclidean topology, give an example of an open set which is not regularly open, and an example of a closed set which is not regularly closed.

1.6. Let $\{H_j\}$ be a family of subsets of the topological space X. Prove or disprove the relations
$$\mathrm{Cl}(\textstyle\bigcup H_j) = \bigcup (\mathrm{Cl}\, H_j), \qquad \mathrm{Cl}(\bigcap H_j) = \bigcap (\mathrm{Cl}\, H_j),$$
in the general case, and do the same when the family is finite. Similarly with the interior operator in place of the closure operator.

1.7. If no proper subset of the topological space X is dense, is the topology necessarily discrete?

1.8. If every countable subset of the topological space X is closed, is the topology necessarily discrete?

1.9. Let X be an infinite set and let x_0 be a point of X. Let $\mathscr{F}$ be the family of subsets of X which are either (i) cofinite or (ii) do not contain x_0. Show that $\mathscr{F}$ is a topology on X in which every point other than x_0 is both open and closed.

1.10. Show that if ϕ is any real-valued injection defined on the set X then a metric ρ on X is given by

$$\rho(\xi, \eta) = |\phi(\xi) - \phi(\eta)|.$$

1.11. Find a topology on the real line, other than the discrete topology and the trivial topology, in which every open set is closed and vice versa.

2
Continuity

2.1 General remarks

In branches of mathematics where the objects of study are sets with some kind of additional structure, a special role is played by functions which preserve that structure. In algebra, for example, where the additional structure takes the form of one or more binary operations, the structure-preserving functions generally called homomorphisms) are those which respect the binary operation or operations.

In topology there are several possible choices for the notion of "structure-preserving" function. Specificially, let $\phi\colon X \to Y$ be a function, where X and Y are topological spaces. Under ϕ, the direct image of a subset H of X is a subset ϕH of Y, while the inverse image of a subset K of Y is a subset $\phi^{-1}K$ of X. At first sight there appear to be four main candidates for the title of structure-preserving function, namely those which satisfy one of the following conditions:

(i) ϕH is open in Y whenever H is open in X,
(ii) ϕH is closed in Y whenever H is closed in X,
(iii) $\phi^{-1}K$ is open in X whenever K is open in Y,
(iv) $\phi^{-1}K$ is closed in X whenever K is closed in Y.

In addition one can obviously consider combinations of these.

It turns out that the direct image conditions (i) and (ii) are less important than the inverse image conditions (iii) and (iv), although just why this is so will emerge only gradually. Certainly, it is partly due to the fact that the inverse image behaves better than the direct image in respect of operations such as complementation.

In fact, (i) and (ii) are distinct conditions, as can easily be seen, while (iii) and (iv) are equivalent conditions. Essentially, therefore, there is just the one inverse image condition, and this is called *continuity*. We study this most important condition in the present chapter and postpone consideration of the direct image conditions until Chapter 4.

2.2 Continuous functions

Definition 2.1

Let $\phi: X \to Y$ be a function, where X and Y are topological spaces. Then ϕ is continuous if the inverse image of each open set of Y is an open set of X or, equivalently, if the inverse image of each closed set of Y is a closed set of X.

Thus ϕ is necessarily continuous if the topology of X is discrete or if the topology of Y is trivial. More interestingly, suppose that X and Y have the cofinite topology. Then ϕ is continuous if and only if $\phi^{-1}(y)$ is finite for each point y of Y.

Constant functions are continuous. For suppose that $\phi X = \{y_0\}$ for some point y_0 of Y. The inverse image of a subset of K of Y is either the full set or the empty set according as K does or does not contain y_0, and so ϕ is continuous.

Note that the identity function on any topological space is continuous. Also that if $\phi: X \to Y$ and $\psi: Y \to Z$ are continuous, where X, Y and Z are topological spaces, then the composition $\psi\phi: X \to Z$ is continuous. This follows immediately from either form of the definition.

Continuity of a function at a point of its domain is defined as follows.

Definition 2.2

Let $\phi: X \to Y$ be a function, where X and Y are topological spaces. Then ϕ is continuous at the point x of X if for each open neighbourhood V of $\phi(x)$ in Y there exists an open neighbourhood U of x in X such that ϕU is contained in V.

In fact ϕ is continuous, as in Definition 2.1, if and only if ϕ is continuous at each point x of X. For suppose that the latter condition is satisfied. If V is open in Y then V is an open neighbourhood of $\phi(x)$ for each point x of $\phi^{-1}V$ and so, by the condition, there exists an open neighbourhood U of x which is contained in $\phi^{-1}V$; thus $\phi^{-1}V$ is open in X.

Conversely, suppose that ϕ is continuous. If x is a point of X and V is an open neighbourhood of $\phi(x)$ in Y then, by continuity of ϕ, $\phi^{-1}Y$ is an open neighbourhood of x in X and, since $\phi\phi^{-1}V \subset V$, this shows that ϕ is continuous at x.

There are various other ways of formulating the continuity condition, such as

Proposition 2.3

Let $\phi\colon X \to Y$ be a function, where X and Y are topological spaces. Then ϕ is continuous if and only if

$$\phi \operatorname{Cl} H \subset \operatorname{Cl} \phi H$$

for each subset H of X.

For suppose that the condition is satisfied for all H, and in particular for $\phi^{-1}F$, where F is closed in Y. Then

$$\phi \operatorname{Cl}(\phi^{-1}F) \subset \operatorname{Cl}(\phi\phi^{-1}F) \subset \operatorname{Cl} F = F,$$

since F is closed. Thus $\operatorname{Cl} \phi^{-1}F \subset \phi^{-1}F$ and so $\phi^{-1}F$ is closed.

Conversely, suppose that ϕ is continuous. Let H be a subset of X and let x be a point of $\operatorname{Cl} H$. If V is an open neighbourhood of $\phi(x)$ in Y then, by continuity, there exists an open neighbourhood U of x such that $\phi U \subset V$. Then ϕU intersects ϕH, since U intersects H, and so V intersects ϕH. Thus $\phi(x) \in \operatorname{Cl} \phi H$. This completes the proof.

Recall that the inverse image behaves well not only in relation to complementation but also in relation to the operations of union and intersection. Thus we obtain

Proposition 2.4

Let $\phi\colon X \to Y$ be a function, where X and Y are topological spaces. Then ϕ is continuous if the inverse image of each member of a generating family for the topology of Y is open in X.

For example, suppose that X and Y are ordered sets and that the function $\phi\colon X \to Y$ is order-preserving, in the sense that

$$\xi < \eta \Rightarrow \phi(\xi) < \phi(\eta) \qquad (\xi, \eta \in X).$$

Also suppose that ϕ is bijective. Then for each point x of X and each open interval (α, β) in Y containing $\phi(x)$ we have the open interval $(\phi^{-1}(\alpha), \phi^{-1}(\beta))$ in X containing x. Since

$$(\phi^{-1}(\alpha), \phi^{-1}(\beta)) = \phi^{-1}(\alpha, \beta)$$

this shows that ϕ is continuous at x, in the interval topology, and so that ϕ is continuous, since x is arbitrary.

In particular, consider the translation function $\phi \colon \mathbb{R} \to \mathbb{R}$ given by $\phi(x) = x + \lambda$, where $\lambda \in \mathbb{R}$. This is bijective and order-preserving, in the standard order, and so continuous in the interval topology.

One can perfectly well use Proposition 2.4 when the codomain is a metric space. However, in that case the following is generally more convenient.

Proposition 2.5

Let $\phi \colon X \to Y$ be a function, where X and Y are topological spaces. Then ϕ is continuous at a point x of X if, in terms of a given neighbourhood base at $\phi(x)$, the inverse image of each basic neighbourhood of $\phi(x)$ contains a neighbourhood of x.

For if V, as in Definition 2.2, is an open neighbourhood of $\phi(x)$ then V contains a basic neighbourhood W of $\phi(x)$. So if $\phi^{-1}W$ contains the neighbourhood U of x then so does $\phi^{-1}V$.

For example, suppose that X and Y are metric spaces with metrics ρ and σ respectively. In this case the condition for $\phi \colon X \to Y$ to be continuous at a given point x of X is that for each positive ε there exists a positive δ such that

$$U_\delta(x) \subset \phi^{-1}V_\varepsilon(\phi(x))$$

or, equivalently, such that

$$\phi U_\delta(x) \subset V_\varepsilon(\phi(x)).$$

Here $U_\delta(x)$ denotes the open δ-ball around x in X and $V_\varepsilon(\phi(x))$ denotes the open ε-ball around $\phi(x)$ in Y.

Clearly, the condition will be satisfied if ϕ is isometric, in the sense that

$$\rho(\xi, \eta) = \sigma(\phi(\xi), \phi(\eta)) \qquad (\xi, \eta \in X),$$

or, more generally, when there exists a positive μ such that

$$\rho(\xi, \eta) = \mu\sigma(\phi(\xi), \phi(\eta)) \qquad (\xi, \eta \in X).$$

In particular, consider the dilatation function $\phi \colon \mathbb{R} \to \mathbb{R}$ given by $\phi(x) = \nu x$, where $\nu \in \mathbb{R}$. We already know that ϕ is continuous when $\nu = 0$. We now see that ϕ is continuous when $\nu \neq 0$, since the above condition is satisfied with $\mu = |\nu|$.

By combining translations and dilatations we obtain the family of affine transformations $\phi \colon \mathbb{R} \to \mathbb{R}$, where

$$\phi(x) = \nu x + \lambda \qquad (\lambda, \nu \in \mathbb{R}).$$

What we have done so far, therefore, shows that affine transformations are continuous. It would be quite possible at this stage to go on to show that real addition and multiplication, as functions $\mathbb{R} \times \mathbb{R} \to \mathbb{R}$, are continuous operations. However, I prefer to develop the theory a little further first.

Returning to the general situation we observe that the continuity conditions can be neatly restated in terms of convergence of filters, as in

Proposition 2.6

Let $\phi \colon X \to Y$ be a function, where X and Y are topological spaces. Then ϕ is continuous at the point x of X if and only if for each filter $\mathscr{F}$ on X which converges to x the direct image filter $\phi_* \mathscr{F}$ on Y converges to $\phi(x)$.

In fact, Definition 2.2 is equivalent to the condition that $\phi_* \mathscr{N}_x$ converges to $\phi(x)$. So if the condition in Proposition 2.6 is satisfied we have only to put $\mathscr{F} = \mathscr{N}_x$ and conclude that ϕ is continuous at x. Conversely, if ϕ is continuous at x and $\mathscr{F}$ converges to x then $\mathscr{F}$ refines $\mathscr{N}_x$, hence $\phi_* \mathscr{F}$ refines $\phi_* \mathscr{N}_x$, hence if $\phi_* \mathscr{N}_x$ converges to $\phi(x)$ then so does $\phi_* \mathscr{F}$.

Corollary 2.7

Let $\phi \colon X \to Y$ be a continuous function, where X and Y are topological spaces. If a sequence $\langle x_n \rangle$ in X converges to a point x then the sequence $\langle \phi(x_n) \rangle$ in Y converges to the point $\phi(x)$.

For an example where the converse of Corollary 2.7 breaks down, consider the real line $\mathbb{R}$ with the "cocountable" topology in which the open sets are the empty set and the complements of the countable subsets. If $\mathbb{Q}'$ is the subset consisting of the irrationals then every rational is obviously an adherence point of $\mathbb{Q}'$ (as in the euclidean topology) although no rational is a limit point of a sequence of irrationals.[5] Thus in $\mathbb{R}$ with the cocountable topology the only convergent sequences are those which are ultimately constant. It follows that the identity function on $\mathbb{R}$, with the cocountable topology, to $\mathbb{R}$, with the euclidean topology, preserves convergent sequences but is not continuous.

---

[5] If the reader has not seen a proof of this I trust it is sufficiently plausible for the purposes of this counterexample.

2.3 Homeomorphisms

Definition 2.8

A homeomorphism $\phi\colon X \to Y$, where X and Y are topological spaces, is a bijective function such that both ϕ and ϕ^{-1} are continuous.

The term *topological equivalence* is often used instead of homeomorphism. Note that it is insufficient for ϕ to be a continuous bijection. For example, the identity function on a set X is continuous when the domain has the discrete topology and the codomain any non-discrete topology but it is not a homeomorphism.

The translation functions $\mathbb{R} \to \mathbb{R}$ are obviously homeomorphisms. More generally, the affine transformation $\phi\colon \mathbb{R} \to \mathbb{R}$ given by

$$\phi(x) = \lambda x + \mu \qquad (\lambda, \mu \in \mathbb{R})$$

is a homeomorphism when $\lambda \neq 0$, since then ϕ is bijective with inverse the affine transformation $\phi^{-1}\colon \mathbb{R} \to \mathbb{R}$ given by

$$\phi^{-1}(x) = \lambda^{-1} x - \lambda^{-1} \mu.$$

Note that if $\phi\colon X' \to X$ is a bijection, where X' is a set and X is a topological space, we can topologize X' so as to make ϕ a homeomorphism. Specifically, we take the open sets of X' to be precisely the inverse images, with respect to ϕ, of the open sets of X. In case the topology of X is generated by a family Γ the topology of X' is generated by the family Γ' consisting of the inverse images of the members of Γ.

For example, consider the set $\mathbb{C}$ of complex numbers. By taking real and imaginary parts we obtain a bijection between $\mathbb{C}$ and the real plane $\mathbb{R}^2$, and then use the above procedure to topologize $\mathbb{C}$. Similarly, in higher dimensions we can topologize $\mathbb{C}^n$ so as to make the corresponding bijection $\mathbb{C}^n \to \mathbb{R}^{2n}$ a homeomorphism.

For another example, consider the set $M(n, \mathbb{R})$ of real $n \times n$ matrices. By listing the entries x_{ij} of each matrix $\|x_{ij}\|$ in a definite order we obtain a bijective correspondence with the real $n \times n$ space, and then topologize $M(n, \mathbb{R})$ so as to make the correspondence a homeomorphism.

Definition 2.9

The topological space X is homogeneous if for each pair of points ξ, η of X there exists a homeomorphism $\phi\colon X \to X$ such that $\phi(\xi) = \eta$.

Discrete spaces and trivial spaces are homogeneous, as are cofinite spaces. Also the real line $\mathbb{R}$ is homogeneous, using translation functions. However, the two-point Sierpinski space is not homogeneous since the function which interchanges the points is not continuous.

2.4 The topological product

In Chapter 1 we have described the properties of the cartesian product $\prod X_j$ of a family $\{X_j\}$ of sets. In the topological situation we shall be particularly concerned with subsets of $\prod X_j$. Among these subsets a special role is played by those which are of the form $\prod H_j$, where H_j is a subset of X_j for each index j; we refer to these as *product sets*. And among the product sets a special role is played by those which are such that $H_j \neq X_j$ for at most a finite number of indices j; we refer to these as *restricted product sets*. In the case of a finite family every product set is restricted, of course, but in the general case the distinction is important.

So now let $\{X_j\}$ be a family of topological spaces. We should like to topologize the cartesian product $\prod X_j$ so that the projection

$$\pi_j \colon \prod X_j \to X_j$$

is continuous, for each index j. This means that the open sets of the topology must include the inverse images $\pi_j^{-1} U_j$, where U_j is open in X_j. The *product topology*, for the cartesian product, is the topology generated by the family of these inverse images, for all indices j and open sets U_j of X_j. The corresponding complete generating family, obtained by taking finite intersections, is precisely the family of restricted product open sets $\prod U_j$, as defined above. With this topology $\prod X_j$ is called the *topological product*, and has the following characteristic property:

Proposition 2.10

Let $\{X_j\}$ be a family of topological spaces. Let $\phi \colon A \to \prod X_j$ be a function, where A is a topological space and $\prod X_j$ is the topological product. Then ϕ is continuous if and only if each of the components $\phi_j = \pi_j \phi \colon A \to X_j$ is continuous.

Naturally, one asks what happens if we take product open sets without restriction as the generating family. The topology on the cartesian product thus defined is called the *box topology*: it is of relatively minor importance. For example, consider the diagonal function

$$\Delta\colon X \to X^J,$$

where X is a topological space. The inverse image of the product open set $\prod U_j$, where U_j is open in X for each index j, is the intersection $\bigcap U_j$. For a restricted product open set the inverse image is open, hence Δ is continuous with the product topology, but Δ is not continuous, in general, with the box topology.

Many of the properties of the topological product can be deduced directly from the characteristic property, without going back to the definition of the product topology. One can see in this way, for example, that the diagonal function

$$\Delta\colon X \to X^J$$

is continuous, since each of its components is the identity function on X. One can also see that if

$$\phi_j\colon X_j \to Y_j \qquad (j \in J)$$

is a family of continuous functions, where X_j and Y_j are topological spaces, then the product

$$\prod \phi_j\colon \prod X_j \to \prod Y_j$$

is also a continuous function. Moreover, if ϕ_j is a homeomorphism for each index j, then so is the product $\prod \phi_j$. It follows, incidentally, that the topological product of the members of a family of homogeneous spaces is a homogeneous space.

To illustrate the general theory let us extend to higher dimensions some of the results already proved for the real line. First, consider the translation function $\phi\colon \mathbb{R}^n \to \mathbb{R}^n$, where

$$\phi(x) = x + \lambda \qquad (\lambda \in \mathbb{R}^n).$$

We recognize this as the product $\prod \phi_j$ of the translation functions $\phi_j\colon \mathbb{R} \to \mathbb{R}$, where

$$\phi_j(x_j) = x_j + \lambda_j \qquad (j = 1, \ldots, n);$$

here λ_j is the jth coordinate of λ. Since each of the ϕ_j is continuous, so is ϕ itself. A similar argument shows that the dilatation function $\phi\colon \mathbb{R}^n \to \mathbb{R}^n$ is continuous, where

$$\phi(x) = \mu x \qquad (\mu \in \mathbb{R}).$$

Returning to the definition of the topological product in general, there are several more points to be made. First of all, the products of closed sets are closed, without restriction, in the product topology. For if E_j is closed in X_j, for each index j, then $\pi_j^{-1}E_j$ is closed in $\prod X_j$ and so the intersection

$$\bigcap \pi_j^{-1}E_j = \prod E_j$$

is closed. More generally, if H_j is a subset of X_j, for each index j, then

$$\text{Cl} \prod H_j = \prod \text{Cl } H_j,$$

without restriction. It should be noted that the corresponding relation for the interior operator is only true for finite products.

It is not always necessary to use the whole family of restricted products of open sets. Thus suppose that the topology of X_j is generated by a family Γ_j of subsets, for each index j. Then the product topology on $\prod X_j$ is generated by the family Γ of restricted products $\prod U_j$, where U_j belongs to Γ_j for each index j for which U_j is not full. Note that Γ generates the product topology if Γ_j generates the topology of X_j for each j.

Another situation in which one can economize on generators arises in relation to the topological Jth power X^J, where X is a topological space. Recall that points of X^J may be interpreted as functions $J \to X$. For each finite subset S of J and each open set U of X let $M(S,U)$ denote the set of functions $\xi: J \to X$ such that $\xi S \subset U$. It is easy to check that the family of subsets of X^J thus defined constitutes a complete generating family for the product topology, in this case. Moreover, if Γ is a complete generating family for the topology of X then the family of subsets $M(S,U)$, for S finite and $U \in \Gamma$, is a complete generating family for the topology of X^J.

It is clear from the definition of the product topology that restricted product neighbourhoods form a neighbourhood base in the topological product. Specifically, let $\{X_j\}$ be a family of topological spaces and let ξ be a point of the topological product $\prod X_j$ with jth projection $\pi_j(\xi) \in X_j$. By a *restricted product neighbourhood* of ξ I mean a neighbourhood of the form $\prod H_j$, where H_j is a neighbourhood of $\pi_j(\xi)$ in X_j and where H_j is full for all but a finite number of indices j. These restricted product neighbourhoods form a neighbourhood base at ξ. More generally, a neighbourhood base at ξ is obtained if the H_j are taken to be members of a neighbourhood base at $\pi_j(\xi)$ for each index j. We deduce

Proposition 2.11

Let $\{X_j\}$ be a family of topological spaces. Then a filter $\mathscr{F}$ on the topological product $\prod X_j$ converges to the point ξ if and only if $\pi_{j*}\mathscr{F}$ converges to $\pi_j(\xi)$ for each index j.

In one direction this is obvious, since the projections π_j are continuous. For the proof in the other direction it is sufficient to show that $\mathscr{F}$ contains each restricted product neighbourhood $\prod H_j$ of ξ. However

$$\prod H_j = \bigcap \pi_j^{-1} H_j,$$

taken over no more than a finite number of indices j. Now $\pi_j^{-1} H_j$ belongs to $\mathscr{F}$, for each such j, since H_j is a neighbourhood of the limit point $\pi_j(\xi)$ of $\pi_{j*}\mathscr{F}$. Therefore the finite intersection belongs to $\mathscr{F}$, which completes the proof.

Corollary 2.12

Let $\{X_j\}$ be a family of topological spaces. Then a sequence $\langle x_n \rangle$ in the topological product $\prod X_j$ converges to the point ξ if and only if the sequence $\langle \pi_j(x_n) \rangle$ converges to $\pi_j(\xi)$ for each index j.

When the indexing set J is finite the proofs of these results can be simplified and further results can be obtained. For example, suppose that each of the X_j is a metric space with metric ρ_j. Since J is finite a metric ρ on the cartesian product $\prod X_j$ is given by

$$\rho(x, x') = \left(\sum \rho_j^2 (\pi_j(x), \pi_j(x')) \right)^{1/2};$$

the formula is modelled on the one used in the euclidean case. Moreover, the metric topology on $\prod X_j$ obtained in this way coincides with the product topology. To see this we have to compare neighbourhood bases at a given point x of $\prod X_j$. Each open ε-ball ($\varepsilon > 0$) at x with respect to ρ contains a product of open $\varepsilon/2$-balls at $\pi_j(x)$ with respect to ρ_j, as illustrated on the left of Figure 2.1. Each product of open ε_j-balls ($\varepsilon_j > 0$) at $\pi_j(x)$ with respect to ρ_j contains an open ε-ball at x with respect to ρ, where

$$\varepsilon = \min_{j \in J} \varepsilon_j,$$

as illustrated on the right of Figure 2.1. So although the neighbourhood bases are different they generate the same neighbourhood filter. Therefore the metric topology agrees with the product topology, as asserted.

A curious result, which is not without its uses, concerns the metric function $\rho \colon X \times X \to \mathbb{R}$ for any metric space X. I assert that ρ is continuous, in the metric topology. For if $\xi, \eta \in X$ and $\varepsilon > 0$ are given, then with $\delta = \varepsilon/2$, we have $|\rho(\xi', \eta') - \rho(\xi, \eta)| < \varepsilon$ whenever $\xi' \in U_\delta(\xi)$ and $\eta' \in U_\delta(\eta)$, which establishes continuity at (ξ, η). A consequence is that the function $X \to \mathbb{R}$, given by $x \mapsto \rho(x, x_0)$ ($x_0 \in X$), is also continuous.

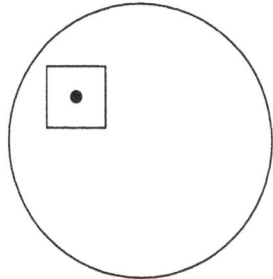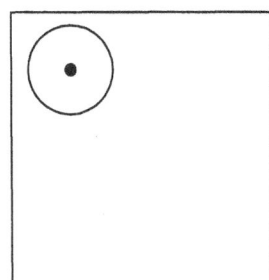

Figure 2.1.

To illustrate the general theory consider the addition function $\psi\colon \mathbb{R} \times \mathbb{R} \to \mathbb{R}$, given by $\psi(s,t) = s + t$. It is simple enough to show that ψ is continuous, of course; one method is as follows. First observe that ψ is continuous at the point $(0,0)$ of the domain. In fact, given the basic open neighbourhood $(-\varepsilon, +\varepsilon)$ of $\psi(0,0) = 0$ in the codomain, where $\varepsilon > 0$, we have

$$\psi((-\delta, +\delta) \times (-\delta, +\delta)) \subset (-\varepsilon, +\varepsilon),$$

where $\delta = \varepsilon/2$. To prove continuity at an arbitrary point (ξ, η) of the domain one can either proceed similarly or else make use of translation functions as follows.

Observe that $\psi(\alpha \times \beta) = \gamma\psi$, as shown below, where α, β, γ are translation by $-\xi, -\eta, -(\xi + \eta)$, respectively.

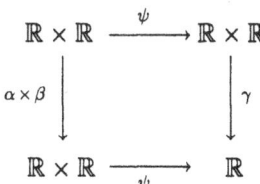

Since α, β are continuous, and ψ is continuous at $(0,0)$, as we have just seen, so $\psi(\alpha \times \beta)$ is continuous at (ξ, η). Therefore $\gamma\psi$ is continuous at (ξ, η), and so ψ is also continuous at (ξ, η), since γ is a homeomorphism.

This mode of procedure may seem more effort than the direct method, in this particular case. However, we shall be using similar procedures in other cases shortly where direct methods are messy and unilluminating.

The argument we have given generalizes at once to higher dimensions and shows that vector space addition $\mathbb{R}^n \times \mathbb{R}^n \to \mathbb{R}^n$ is continuous, for all n. Hence vector space addition $\mathbb{C}^n \times \mathbb{C}^n \to \mathbb{C}^n$ is continuous, for all n. Also matrix addition

$$M(n, \mathbb{R}) \times M(n, \mathbb{R}) \to M(n, \mathbb{R})$$

is continuous, for all n.

Now let us turn to the multiplication function $\phi \colon \mathbb{R} \times \mathbb{R} \to \mathbb{R}$. To show that ϕ is continuous at the point $(0,0)$ of the domain, recall that a neighbourhood base for the point $\phi(0,0) = 0$ of the codomain is formed by the open intervals $(-\varepsilon, +\varepsilon)$, where $0 < \varepsilon < 1$. Since

$$\phi((-\varepsilon, +\varepsilon) \times (-\varepsilon, +\varepsilon)) \subset (-\varepsilon, +\varepsilon),$$

continuity at $(0,0)$ follows at once. To show that ϕ is continuous at the point $(1,1)$ of the domain, recall that a neighbourhood base for the point $\phi(1,1) = 1$ of the codomain is formed by the open intervals $(1 - 2\varepsilon, 1 + 3\varepsilon)$, where $0 < \varepsilon < 1$. Since $\phi((1 - \varepsilon, 1 + \varepsilon) \times (1 - \varepsilon, 1 + \varepsilon)) \subset (1 - 2\varepsilon, 1 + 3\varepsilon)$, continuity at (1,1) follows at once. To show that ϕ is continuous at the point $(1,0)$ of the domain, observe that a neighbourhood base at the point $\phi(1,0) = 0$ of the codomain is formed by the open intervals $(-2\varepsilon, 2\varepsilon)$, where $0 < \varepsilon < 1$. Since

$$\phi((1 - \varepsilon, 1 + \varepsilon) \times (-\varepsilon, +\varepsilon)) \subset (-2\varepsilon, +2\varepsilon),$$

continuity at $(1,0)$ follows at once, and continuity at $(0,1)$ follows by symmetry.

So now consider the general case. Given the point (ξ, η) of $\mathbb{R} \times \mathbb{R}$, we have a commutative diagram as shown below, where α denotes dilatation by ξ for $\xi \neq 0$, the identity for $\xi = 0$, where β denotes dilatation by η for $\eta \neq 0$, the identity for $\eta = 0$, and where $\gamma = \alpha\beta$.

$$
\begin{array}{ccc}
\mathbb{R} \times \mathbb{R} & \xrightarrow{\;\alpha \times \beta\;} & \mathbb{R} \times \mathbb{R} \\[1mm]
\phi \downarrow & & \downarrow \phi \\[2mm]
\mathbb{R} & \xrightarrow[\;\gamma\;]{} & \mathbb{R}
\end{array}
$$

Now α, β, γ are homeomorphisms in all cases, as we have seen. If ξ, η are non-zero then ϕ on the left is continuous at $(1,1)$, as we have shown, so $\gamma\phi = \phi(\alpha \times \beta)$ is continuous at $(1,1)$, so ϕ is continuous at (ξ, η). If ξ is non-zero and $\eta = 0$ we argue in the same way, replacing $(1,1)$ by $(1,0)$ and similarly if $\xi = 0$ and $\eta \neq 0$. Thus all cases are covered and we see that real multiplication is continuous.

Similarly in higher dimensions. To show that scalar multiplication

$$\theta \colon \mathbb{R} \times \mathbb{R}^n \to \mathbb{R}^n$$

is continuous we use the above and Proposition 2.10. Specifically, to show that the jth component $\theta_j = \pi_j\theta$ of θ is continuous $(j = 1, \ldots, n)$ we observe that $\theta_j = \phi(\mathrm{id} \times \pi_j)$, where $\phi \colon \mathbb{R} \times \mathbb{R} \to \mathbb{R}$ is real multiplication, as before.

A somewhat similar argument shows that bilinear functions $\mathbb{R}^p \times \mathbb{R}^q \to \mathbb{R}$ are continuous, for all p, q, and then again that matrix multiplication

$$M(n, \mathbb{R}) \times M(n, \mathbb{R}) \to M(n, \mathbb{R})$$

is continuous, since each of the components is bilinear. A further example is the determinant function

$$\det\colon M(n,\mathbb{R}) \to \mathbb{R}.$$

Similarly in the complex case.

Of course, one is not restricted to the linear situation. For example, to show that the squaring function $\mathbb{R} \to \mathbb{R}$, given by $x \mapsto x^2$, is continuous, one expresses the function as a composition

$$\mathbb{R} \xrightarrow{\Delta} \mathbb{R} \times \mathbb{R} \xrightarrow{\phi} \mathbb{R}.$$

More generally, by combining the operations of addition and multiplication, in obvious ways, one obtains that each polynomial function $\mathbb{R} \to \mathbb{R}$ is continuous, where

$$x \mapsto a_0 + a_1 x + \cdots + a_n x^n \qquad (a_i \in \mathbb{R}).$$

Similarly for complex numbers.

Returning to the general case there are some remarks to be made on the subject of relations. First of all, observe that a relation R on the topological space X is, of course, a subset of the topological product $X \times X$. When R is an equivalence relation, the equivalence classes are open in X when R is open, and closed in X when R is closed. Not quite so obviously the equivalence classes are also closed in X when R is open, because the complement of one equivalence class is just the union of all the others, and the union of open sets is open. In the literature, incidentally, the terms open relation and closed relation are used in a different sense.

In this connection some interesting questions arise concerning neighbourhoods of the diagonal ΔX in the topological product $X \times X$. Such neighbourhoods can be regarded as relations on X and so the terminology and notation set forth in the preliminary chapter are applicable. All such neighbourhoods are reflexive, of course, since they contain the diagonal. Neighbourhoods of the special form

$$\bigcup (U_j \times U_j),$$

where $\{U_j\}$ is an open covering of X, are also symmetric. When the members of the covering are mutually disjoint such neighbourhoods are also transitive.

Every neighbourhood U of the diagonal contains the symmetric neighbourhood $U \cap U^{-1}$. However, such a neighbourhood does not necessarily contain a transitive neighbourhood, or even a neighbourhood V of the diagonal which satisfies the weaker condition $V \circ V \subset U$. We shall pursue these questions further in Chapter 8.

2.5 Topological groups

Let us turn now to a different subject. The notion of topological group can be introduced quite early in the study of topology, as soon as the topological product has been defined. I assume that the reader is already familiar with the axioms of group theory in the algebraic sense. So let G be a group in that sense. Then G is equipped with a neutral element e, a symmetry $u\colon G \to G$, and a binary operation $m\colon G \times G \to G$. Usually the binary operation is called multiplication, in which case the symmetry is called inversion and the neutral element the identity; however, in some situations it seems more natural to call the binary operation addition, in which case the symmetry is called the negative and the neutral element the zero. Normally we use the multiplicative terminology. Suppose that G, and hence $G \times G$, has a topology. If the functions u and m are continuous then we describe G as a *topological group*. For example, if the topology of G is discrete or trivial then the functions are continuous and the group G is topological.

From what we have already proved the real line $\mathbb{R}$ is a topological group, with the additive group structure and the euclidean topology, and more generally the real n-space $\mathbb{R}^n$ is a topological group similarly. Further examples will be given in the next chapter.

In the theory of topological groups homomorphisms are required to be continuous, as well as satisfying the usual algebraic requirements. Thus a function $\phi\colon G \to H$, where G and H are topological groups, is a homomorphism in the topological sense if it is both a homomorphism in the algebraic sense and a continuous function. Moreover ϕ is an isomorphism in the topological sense if it is both an isomorphism in the algebraic sense and a homeomorphism.

Due to the interplay between the algebra and the topology a topological group has many special topological properties such as homogeneity. I conclude this chapter with a brief discussion of these which will also serve to illustrate various points in the theory we hve developed so far.

Let G be a topological group. For each element $g \in G$ a function $g_\#\colon G \to G$ is defined by multiplying on the left by g. This is known as *left translation*; right translation is defined similarly. The continuity of $g_\#$ can be established by expressing it as the composition

$$G \xrightarrow{\ \Delta\ } G \times G \xrightarrow{\ c \times \mathrm{id}\ } G \times G \xrightarrow{\ m\ } G,$$

where c is constant at g. Since $g_\#$ and $(g^{-1})_\#$ are continuous, and since $(g^{-1})_\#$ is the (set-theoretic) inverse of $g_\#$ we conclude that $g_\#$ is a homeomorphism. Thus left translation is a homeomorphism, and similarly so is right translation; combining the two we see that conjugation is also a homeomorphism.

Since translations are homeomorphisms it follows at once that G is a homogeneous space. Another useful consequence is

Proposition 2.13

Let $\phi\colon G \to H$ be a homomorphism in the algebraic sense, where G and H are topological groups. If ϕ is continuous at the neutral element of G then ϕ is continuous everywhere.

For let g be any element of G, and let V be an open neighbourhood of $\phi(g)$. Then $(\phi g)_\#^{-1} V$ is an open neighbourhood of $\phi(e)$, where e is the neutral element of G. Let U be an open neighbourhood of e in G such that $\phi U \subset (\phi g)_\#^{-1} V$. Then $g_\# U$ is an open neighbourhood of g in G such that $\phi(g_\# U) \subset V$. Thus ϕ is continuous at g, and so the result is obtained.

If A is a subset of the topological group G we denote by A^{-1} the direct image of A under the inversion $G \to G$. Clearly A^{-1} is open whenever A is open. Similarly, if A, B are subsets of G we denote by $A \cdot B$ the direct image of $A \times B$ under the multiplication $G \times G \to G$, with the usual modification for one-point subsets. It is important to note that $A \cdot B$ is open in G whenever A or B is open in G. For take $B = U$, where U is open in G. Each of the translates $g_\# U$ of U is open in G, since $g_\# U$ is the inverse image of U under the translation homeomorphism $g_\#^{-1}$. Therefore the union of these translates, as g runs through the elements of A, is also open in G. But this is precisely $A \cdot U$. Similarly, $U \cdot A$ is open in G.

If U is a neighbourhood of the neutral element e then U^{-1} is also a neighbourhood. Hence U contains the neighbourhood $V = U \cap U^{-1}$ which is symmetric in the sense that $V = V^{-1}$. Again, using continuity of multiplication, we see that $m^{-1}U$ contains a product neighbourhood $W_1 \times W_2$ of (e, e) in $G \times G$ and so U contains the neighbourhood $W = W_1 \cap W_2$ which satisfies the condition $W \cdot W \subset U$. Neighbourhood bases whose members satisfy these and other special requirements are frequently used in the theory of topological groups.

Finally, we prove two results concerning subgroups: at this stage we simply use the term subgroup in the algebraic sense.

Example 2.14

Let H be a subgroup of the topological group G. If H is open in G then H is closed in G. If H is closed in G and of finite index in G then H is open in G.

To see this observe that $G - H$ is a union of cosets of H. If H is open then each of the cosets is open, by translation, hence $G - H$ is open and H is closed. If H is closed then each of the cosets is closed, by translation. If, further, the index is finite then the cosets are finite in number, hence $G - H$ is closed and so H is open.

Example 2.15

Let H be a subgroup of the topological group G. Then the closure $\mathrm{Cl}\, H$ of H is also a subgroup of G. Moreover, $\mathrm{Cl}\, H$ is normal if H is normal.

For the first assertion consider the division function $d \colon G \times G \to G$. Since d is continuous it follows from Proposition 2.3 that

$$\mathrm{Cl}\, A \cdot \mathrm{Cl}(B^{-1}) \subset \mathrm{Cl}(A \cdot B^{-1})$$

for any subsets A, B of G, from which the first assertion follows at once. For the second consider the conjugation function $\gamma \colon G \to G$ determined by a given element g of G. Conjugation is continuous, as we have seen. By Proposition 2.3 again we have that

$$\gamma(\mathrm{Cl}\, A) \subset \mathrm{Cl}(\gamma A)$$

for any subset A of G, from which the second assertion follows at once.

For example take $A = \{e\}$: we see that $\mathrm{Cl}\,\{e\}$ is a normal subgroup.

If $\{G_j\}$ is a family of topological groups then the cartesian product $\prod G_j$ is also a topological group, with the direct product group structure and the product topology. This is a simple consequence of Proposition 2.10, which also shows that each of the projections

$$\pi_j \colon \prod G_j \to G_j$$

is a (continuous) homomorphism. Moreover, a function

$$\phi \colon G \to \prod G_j,$$

where G is a topological group, is a homomorphism if and only if each of the functions ϕ_j is a homomorphism. It follows, for example, that the product of homomorphisms is again a homomorphism. It also follows that the diagonal function

$$\Delta \colon G \to G^J$$

is a homomorphism, for each topological group G and indexing set J.

Exercises

2.1. Let $\{p_j\}$ be a family of polynomials in the real variable t with real coefficients. Show that the set of values of t where $p_j(t) = 0$ for all j is a closed set of $\mathbb{R}$, with the euclidean topology.

2.2. Let $E = \{0, 1\}$ be the Sierpinski space with $\{0\}$ open and $\{1\}$ not. Which are the closed sets of $E \times E$?

2.3. Show that the topological product of an infinite number of discrete spaces is not in general discrete.

2.4. Show that the direct product of a family of topological groups is a topological group not only with the product topology but also with the box topology.

2.5. Suppose that the neutral element e forms a closed set in the topological group G. Show that the diagonal subset ΔG forms a closed subgroup of the direct product $G \times G$.

2.6. Show that each closed subgroup of the real line $\mathbb{R}$ with euclidean topology is either the full group $\mathbb{R}$ or else is the subgroup consisting of integral multiples of α for some positive real number α.

2.7. Show that if X is an infinite set with the cofinite topology then every continuous real-valued function on X is constant.

2.8. Show that in a topological group G, the open symmetric neighbourhoods of the neutral element e form a base for the neighbourhood filter at e.

2.9. Show that any group G can be made into a topological group by specifying, as neighbourhood base at the neutral element, e, a non-empty family Γ of subgroups of G such that

(i) if $U, V \in \Gamma$ then $W \subset U \cap V$ for some $W \in \Gamma$,
(ii) if $U \in \Gamma$ and $g \in G$ then $g^{-1}Vg \subset U$ for some $V \in \Gamma$.

Illustrate by taking Γ to be

(a) the members of a chain of normal subgroups,
(b) the family of subgroups of finite index,
(c) the members of the lower central series.

2.10. Show that the closure $\mathrm{Cl}\, E$ of a subset E of the topological group G coincides with the intersection of the subsets $N \cdot E$, where N runs through the open neighbourhoods of the neutral element e.

2.11. Let U be an open neighbourhood of the neutral element e in the topological group G. Show that the set of elements $g \in G$ such that $g^2 \in U$ coincides with the union of the open V such that $V^2 \subset U$.

2.12. Show that every subgroup H of the topological group G with interior Int H non-empty is both open and closed.

2.13. Let G be a topological group, let H be a dense subgroup of G, and let K be a normal subgroup of H. Show that Cl K is a normal subgroup of G.

2.14. Suppose that the neutral element e forms a closed set in the topological group G. Show that the centre of G is a closed normal subgroup of G.

2.15. Let G be a topological group. Show that

  (i) the intersection of all open subgroups of G is a normal subgroup of G, and that
  (ii) the intersection of all non-trivial closed subgroups of G is a normal subgroup of G.

The Induced Topology and Its Dual

3.1 The induced topology

This chapter is mainly concerned with subspaces and quotient spaces. However, it often happens in mathematics that by taking a more general point of view one can see more clearly what is happening in a special case. We begin, therefore, by discussing the notion of induced topology before going on to embeddings and subspaces; likewise, we discuss the notion of coinduced topology before going on to quotient maps and quotient spaces.[6]

Definition 3.1

Let $\phi: X \to Y$ be a function, where X and Y are topological spaces. The topology of X is induced by ϕ from the topology of Y if the open sets of X are precisely the inverse images, with respect to ϕ, of the open sets of Y.

The condition here is stronger than continuity, since X is not allowed to have open sets other than those which arise from Y in the prescribed manner. Because of the good behaviour of the inverse image with respect to complementation we see at once that "open set" may be replaced by "closed set" in both places without affecting the meaning of the definition.

---

[6] The induced and coinduced topologies are also known as the initial and final topologies, respectively.

We have already observed that the inverse image behaves well in relation to the operations of union and intersection. This implies that the condition in Definition 3.1 only needs to be checked for the members of a generating family for the topology of the domain, as stated in

Proposition 3.2

Let $\phi\colon X \to Y$ be continuous, where X and Y are topological spaces. Suppose that each member of a generating family for the topology of X is the inverse image of an open set of Y. Then the topology of X is induced by ϕ from the topology of Y.

The situation may arise in which we are given a function $\phi\colon X \to Y$, where Y is a topological space and X is a set. Then we can use the procedure indicated in Definition 3.1 to give X a topology. This topology, which is called the *induced topology*, may be described as the coarsest topology such that ϕ is continuous. For example, Y has the trivial topology then so does X, under this procedure.

The induced topology is transitive in the following sense. Let $\phi\colon X \to Y$ and $\psi\colon Y \to Z$ be functions, where X, Y and Z are topological spaces. If Y has the topology induced by ψ from the topology of Z and X has the topology induced by ϕ from the topology of Y then X has the topology induced by $\psi\phi$ from the topology of Z. The proof is obvious.

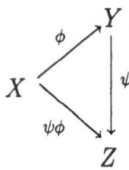

The following property is characteristic of the induced topology.

Proposition 3.3

Let $\phi\colon X \to Y$ and $\psi\colon Y \to Z$ be functions, where X, Y and Z are topological spaces. Suppose that Y has the topology induced by ψ from the topology of Z. Then ϕ is continuous if (and only if) $\psi\phi$ is continuous.

For if V is open in Y then $V = \psi^{-1}W$, where W is open in Z, since Y has the induced topology, and then

$$\phi^{-1}V = \phi^{-1}\psi^{-1}W = (\psi\phi)^{-1}W$$

is open in X, since $\psi\phi$ is continuous.

Multiple forms of the induced topology can also be considered. Thus let X be a set and let $\{\phi_j\}$ be a family of functions $\phi_j \colon X \to Y_j$, where Y_j is a topological space, for each index j. We can give X the coarsest topology which makes ϕ_j continuous for each j. Thus a generating family consists of the inverse images $\phi_j^{-1} U_j$, where j runs through the indices and U_j runs through the open sets of Y_j. The topological product is an example of this, with each function ϕ_j the corresponding projection of the cartesian product. In fact this observation implies that the multiple form of the induced topology, as above, is no more than the ordinary form of the induced topology with respect to the function

$$\phi \colon X \to \prod Y_j$$

with jth component ϕ_j for each index j. For this reason it is not necessary to say very much about the multiple form.

Returning to the ordinary form, consider the situation where the function in question is injective. Because of the importance of this case special terminology is used.

Definition 3.4

Let $\phi \colon X \to Y$ be an injection, where X and Y are topological spaces. If X has the topology induced by ϕ from the topology of Y then ϕ is a (topological) embedding.

A continuous injection is necessarily an embedding when the domain has trivial topology, for example, when the domain is a one-point space. However, in general, the domain of a continuous injection will have open sets and closed sets which are not inverse images of open sets and closed sets of the codomain. For example, take the identity function on any set with the topology of the domain a strict refinement of the topology of the codomain.

Of course, Propositions 3.2 and 3.3 and other results about the induced topology apply to embeddings as a special case. In addition, we have

Proposition 3.5

Let $\phi \colon X \to Y$ and $\psi \colon Y \to Z$ be continuous functions, where X, Y and Z are topological spaces. If $\psi\phi$ is an embedding then so is ϕ.

For then each open set of X is of the form $(\psi\phi)^{-1} W$, where W is open in Z, and so is of the form $\phi^{-1}\psi^{-1} W$, where $\psi^{-1} W$ is open in Y.

In particular, take $Z = X$ and $\psi\phi$ the identity. In that case ψ is called a *left inverse* of ϕ and Proposition 3.5 states that a continuous function is an embedding if it admits a (continuous) left inverse. Thus we obtain

Corollary 3.6

Let $\phi: X \to Y$ be a continuous function, where X and Y are topological spaces. Then the graph function

$$\Gamma_\phi: X \to X \times Y$$

is an embedding.

Here Γ_ϕ may be expressed as the composition

$$X \xrightarrow{\Delta} X \times X \xrightarrow{\mathrm{id} \times \phi} X \times Y;$$

the left inverse is given by projecting onto the first factor. Similarly, we obtain

Corollary 3.7

For any topological space X the diagonal function

$$\Delta: X \to X^J$$

is an embedding.

Another straightforward application of Proposition 3.2 is

Proposition 3.8

Let $\{\phi_j\}$ be a family of embeddings $\phi_j: X_j \to Y_j$, where X_j and Y_j are topological spaces. Then the product

$$\prod \phi_j: \prod X_j \to \prod Y_j$$

is an embedding.

Given topological spaces A and X one may ask whether A can be embedded in X, i.e. whether a function $\sigma: A \to X$ exists which satisfies the embedding condition. This is not an easy question, in general, although there are some cases where it can be answered by elementary methods. For example, it is obviously impossible to embed A in X if X has the discrete topology, or the trivial topology, and A does not.

3.2 Subspaces

Suppose now that we have a topological space X and a subset A of X. For each subset H of X the trace $A \cap H$ of H on A is just the inverse image $\sigma^{-1}H$ of H under the injection $\sigma: A \to X$, as illustrated in Figure 3.1. We use the induced topology to topologize A so as to make σ an embedding. Specifically, we take the open sets of A to be the traces on A of the open sets of X or, equivalently, the closed sets of A to be the traces on A of the closed sets of X. This is usually called the *relative topology* and A, with this topology, is called a *subspace* of X. In fact the terms subset and subspace tend to be used interchangeably since it is most unusual to give a subset any topology other than the relative topology.

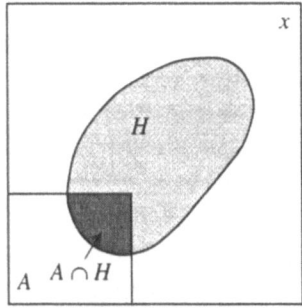

Figure 3.1.

It is important to appreciate that, in the above situation, a subset H of A itself may be open in A but not open in X, or closed in A but not closed in X. For example, A is always open and closed in itself but this says nothing about its status in relation to X.

The composition of embeddings is an embedding, as we have seen. Hence, if A is a subspace of X and B is a subspace of A then B is a subspace of X. Notice that a generating family for the topology of X determines, by taking traces, a generating family for the relative topology of each subset A of X.

Some of the results we have obtained for embeddings are particularly useful in relation to subspaces. Thus Proposition 3.3 shows that if $\phi: X \to Y$ is continuous, where X and Y are topological spaces, then so is the function $\phi': X' \to Y'$ determined by ϕ for each subspace X' of X and each subspace Y' of Y which contains $\phi X'$. Moreover, Proposition 3.5 shows that if, further, ϕ is an embedding then so is ϕ'. Consequently, an embedding $\phi: X \to Y$ maps X homeomorphically onto the subspace ϕX of Y; more generally, ϕ maps X' homeomorphically onto $\phi X'$ for every subspace X' of X. For example, take $X = Y = \mathbb{R}$, with the euclidean topology. Choosing ϕ to be an affine transformation we see that all the open intervals (α, β) $(\alpha < \beta)$ are homeomorphic, and that all the closed intervals $[\alpha, \beta]$ are homeomorphic. We also see that all the half-open intervals $[\alpha, \beta)$ and $(\beta, \alpha]$ are homeomorphic.

For another example, recall that the (upper) Sorgenfrey line L is just the real line with the topology generated by intervals of the form $[\alpha, \beta)$. The Sorgenfrey plane $L \times L$ is the topological product of L with itself. The topology of $L \times L$ is therefore generated by slabs of the form

$$[\alpha_1, \beta_1) \times [\alpha_2, \beta_2).$$

As we have seen in Corollary 3.7, the diagonal subspace $\{(\xi, \eta) : \xi = \eta\}$ is homeomorphic to L, on general grounds. However, the antidiagonal subspace $\{(\xi, \eta) : \xi = -\eta\}$ is discrete, in the relative topology, since each point $(\xi, -\xi)$ is the trace on the antidiagonal of the open slab $[\xi, \infty) \times [-\xi, \infty)$ and therefore open in the relative topology.

The observation about generating families shows that the relative topology is compatible with the topological product. Specifically, let $\{X_j\}$ be a family of topological spaces and let A_j be a subset of X_j $(j \in J)$. We can first give each A_j the relative topology and then give $\prod A_j$ the product topology, or we can first give $\prod X_j$ the product topology and then give $\prod A_j$ the relative topology; the result is exactly the same, by Proposition 3.8.

To illustrate this consider a subgroup H of a topological group G. I claim that H, with the relative topology, is also a group in the topological sense. This can easily be seen from the following diagrams, where m, n are the binary operations, where u, v are the inversions and where σ is the embedding.

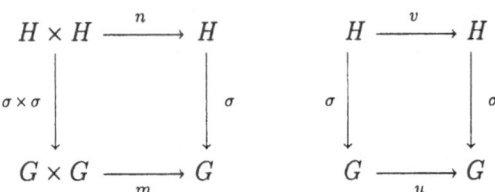

From now on we will always regard a subgroup of a topological group as being itself a topological group, in this manner.

Returning to the case where A is a subset of the topological space X, with the relative topology, observe that for a point x of A, the trace on A of the neighbourhood filter of x in X is a base for the neighbourhood filter of x in A; more generally, the trace of a neighbourhood base of x in X is a base for the neighbourhood filter of x in A. It follows that if X is a metric space then the relative topology which A obtains from the metric topology on X coincides with the metric topology determined by the restriction to $A \times A$ of the metric on X.

Before giving some applications and examples there is a further theoretical result which needs to be discussed. Recall that a *covering* of a set X is a family of subsets of X such that each point of X belongs to at least one member of the family. A subfamily of a covering is called a *subcovering* if it too is a covering. If A is a subset of X we can consider coverings of A by families of subsets of X, in the obvious sense.

When X is a topological space the types of covering which are important are those which consist entirely of open sets or entirely of closed sets. These are called *open coverings* or *closed coverings*, as the case may be, and we have

Proposition 3.9

Let $\{A_j\}$ be a family of subsets of the topological space X which is either an open covering or a finite closed covering of X. Let H be a subset of X and let $H_j = A_j \cap H$, for each index j. Then H is open in X if H_j is open in A_j for each j, while H is closed in X if H_j is closed in A_j for each j.

First, suppose that $\{A_j\}$ is an open covering. If H_j is open in A_j for each j then H_j is open in X for each j and so the union H of the H_j is open in X. If H_j is closed in A_j for each j, replace H by $X - H$ and proceed as before.

Second, suppose instead that $\{A_j\}$ is a finite closed covering. If H_j is closed in A_j for each j then H_j is closed in X for each j and so the union H of the H_j is closed in X. If H_j is open in A_j for each j, replace X by $X - H$ and proceed as before.

Corollary 3.10

Let $\{A_j\}$ be a family of subsets of the topological space X which is either an open covering or a finite closed covering. Then a function $\phi \colon X \to Y$, for any topological space Y, is continuous if (and only if) the restriction $\phi | A_j$ is continuous for each index j.

For a simple application, consider the function

$$\min \colon \mathbb{R} \times \mathbb{R} \to \mathbb{R}.$$

The subsets

$$H_+ = \{(\xi, \eta) \colon \xi \geq \eta\}, \qquad H_- = \{(\xi, \eta) \colon \xi \leq \eta\}$$

are closed and cover $\mathbb{R} \times \mathbb{R}$. Now min is given by the second projection on H_+, the first projection on H_-. Since the projections are continuous the corollary shows that min is continuous. Similarly max is continuous.

To illustrate the ideas we have been discussing let us begin with subsets of the real line $\mathbb{R}$, with the euclidean topology. Given a subset A, a generating family for the relative topology consists of the traces $A \cap (\alpha, \beta)$ of the open intervals of $\mathbb{R}$, as we have seen. In particular cases this may not be the most convenient family to use. For example, the relative topology on the set $\mathbb{Z}$ of integers is discrete, since each integer n has the open neighbourhood $(n-1, n+1)$ in $\mathbb{R}$ of which the trace is $\{n\}$; in this case the obvious generating family to use is that formed by the subsets $\{n\}$ for each integer n. For another example, the relative topology on the rational line $\mathbb{Q}$ is generated by the rational intervals

$$(\alpha, \beta) = \{\xi \in \mathbb{Q} : \alpha < \xi < \beta\},$$

for rational α, β. Note that for real open intervals with irrational end-points the rational trace is both open and closed. Similar observations apply in the case of subsets of the real n-space $\mathbb{R}^n$, with the euclidean topology.

Next consider the punctured real line $\mathbb{R} - \{0\} = \mathbb{R}_*$ with the relative topology. Note that the trace of the open interval (α, β) is (α, β) itself unless $\alpha < 0 < \beta$ when the trace is the union of the open intervals $(\alpha, 0)$ and $(0, \beta)$. I assert that the inversion function $u \colon \mathbb{R}_* \to \mathbb{R}_*$ is continuous, where $u(\xi) = \xi^{-1} (\xi \in \mathbb{R}_*)$. For consider the trace of the open interval (α, β). If both α and β are non-zero the inverse image of the trace is the trace of $(\alpha^{-1}, \beta^{-1})$ or $(\beta^{-1}, \alpha^{-1})$. If α is non-zero the inverse image of $(\alpha, 0)$ is $(\alpha^{-1}, 0)$, while if β is non-zero the inverse image of $(0, \beta)$ is $(0, \beta^{-1})$. Thus, in any case, the inverse image is open and so u is continuous, as asserted.

We have already seen that the multiplication function $\mathbb{R} \times \mathbb{R} \to \mathbb{R}$ is continuous, hence the multiplication function $\mathbb{R}_* \times \mathbb{R}_* \to \mathbb{R}_*$ is continuous. Since inversion is also continuous, as we have just shown, we conclude that $\mathbb{R}_*$ is a topological group. Another consequence is that the multiplication function $\mathbb{R} \times \mathbb{R}_* \to \mathbb{R}$ is continuous and hence that the division function $\mathbb{R} \times \mathbb{R}_* \to \mathbb{R}$ is continuous. These results imply that rational functions $\mathbb{R} \to \mathbb{R}$, with non-vanishing denominator, are continuous.

Similar results hold for complex numbers. Thus the punctured complex line $\mathbb{C}_* = \mathbb{C} - \{0\}$ forms a topological group, under complex multiplication, and the complex numbers of unit modulus form a subgroup, called the circle group.

This is a convenient point at which to mention the well-known function $\alpha \colon \mathbb{R} \times \mathbb{R} \to \mathbb{R}$ given by

$$\alpha(\xi, \eta) = \frac{2\xi\eta}{\xi^2 + \eta^2}$$

away from $(0,0)$ and by $\alpha(0,0) = 0$. If we fix either ξ or η the resulting function $\mathbb{R} \to \mathbb{R}$ is continuous. However, if we compose α with the diagonal function Δ the result is not continuous, hence α itself is not continuous.

Turning now to higher dimensions we recall that scalar multiplication

$$\mathbb{R}^n \times \mathbb{R} \to \mathbb{R}^n$$

is continuous, hence the restriction

$$\mathbb{R}^n \times \mathbb{R}_* \to \mathbb{R}^n$$

is continuous. Composing the restriction with the product of the identity on $\mathbb{R}^n$ and the inversion function $\mathbb{R}_* \to \mathbb{R}_*$ we deduce that scalar division

$$\mathbb{R}^n \times \mathbb{R}_* \to \mathbb{R}^n$$

is continuous.

By way of application let us show that the real n-space $\mathbb{R}^n$ is homeomorphic to the open n-ball U^n. In fact $\phi: \mathbb{R}^n \to U^n$ is a continuous function with continuous inverse $\psi: U^n \to \mathbb{R}^n$, where

$$\phi(\xi) = \xi(1 + \|\xi\|)^{-1} \qquad (\xi \in \mathbb{R}^n),$$

$$\psi(\eta) = \eta(1 - \|\eta\|)^{-1} \qquad (\eta \in U^n).$$

For another example of the same sort consider the n-sphere S^n in $\mathbb{R}^{n+1}$. Let $p \in S^n$ be the pole where the last coordinate $x_{n+1} = 1$. Stereographic projection from p onto the equatorial n-space determines a function $\phi: S^n - \{p\} \to \mathbb{R}^n$, as indicated in Figure 3.2, where

$$\phi(x_1, \ldots, x_{n+1}) = \left(\frac{x_1}{1 - x_{n+1}}, \ldots, \frac{x_n}{1 - x_{n+1}} \right),$$

which is readily shown to be a homeomorphism.

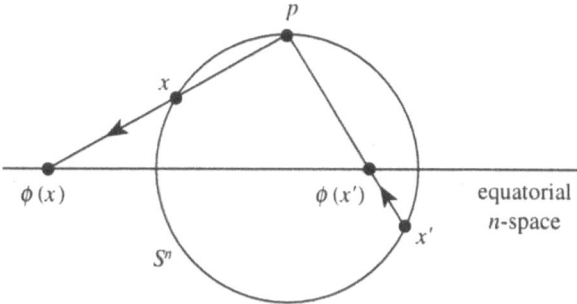

Fig. 3.2

For an example of a different type consider the general linear group $\mathrm{Gl}(n, \mathbb{R})$ of non-singular real $n \times n$ matrices, where $n = 0, 1, \ldots$; the group operation is matrix multiplication. I assert that $\mathrm{Gl}(n, R)$, topologized as a subset of $M(n, \mathbb{R})$, is a group in the topological sense. For the group operation is continuous since matrix multiplication

$$M(n, \mathbb{R}) \times M(n, \mathbb{R}) \to M(n, \mathbb{R})$$

is continuous, as we saw in the previous chapter. Also the inversion function is continuous since it can be expressed as the result of taking the matrix of co-factors, then transposing and finally dividing by the determinant, all of which are continuous operations. Thus $\mathrm{Gl}(n, \mathbb{R})$ is a topological group, as asserted.

3.3 The coinduced topology

We now turn from the induced topology to the coinduced topology, which is in some sense a dual notion.

Definition 3.11

Let $\phi: X \to Y$ be a function, where X and Y are topological spaces. The topology of Y is coinduced by ϕ from the topology of X if the open sets of Y are precisely those subsets of which the inverse images, with respect to ϕ, are open sets of X.

The condition here is stronger than continuity; when ϕ is continuous there will generally be non-open subsets of Y of which the inverse images are open sets of X. Because of the good behaviour of the inverse image with respect to comple-mentation we see at once that "open sets" may be replaced by "closed sets" in both places without affecting the meaning of the definition.

The situation may arise in which we are given a function $\phi: X \to Y$, where X is a topological space and Y is a set. Then we can use the procedure indicated in Definition 3.11 to give Y a topology. This topology, which is called the *coinduced topology*, may be described as the finest topology such that ϕ is continuous. For example, if X has the discrete topology then so does Y, under this procedure.

The coinduced topology is transitive in the following sense. Let $\phi: X \to Y$ and $\psi: Y \to Z$ be functions, where X, Y and Z are topological spaces. If Y has the topology coinduced by ϕ from the topology of X and Z has the topology coinduced by ψ from the topology of Y then Z has the topology coinduced by $\psi\phi$ from the topology of X.

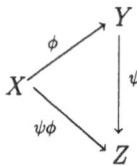

The following property is characteristic of the coinduced topology.

Proposition 3.12

Let $\phi: X \to Y$ and $\psi: Y \to Z$ be functions, where X, Y and Z are topological spaces. Suppose that Y has the topology coinduced by ϕ from the topology of X. Then ψ is continuous if (and only if) $\psi\phi$ is continuous.

For if W is open in Z then $(\psi\phi)^{-1}W$ is open in X, since ψ is continuous, hence $\psi^{-1}W$ is open in Y, since Y has the coinduced topology.

The case when ϕ is surjective is particularly important, and special terminology is used.

Definition 3.13

Let $\phi: X \to Y$ be a surjection, where X and Y are topological spaces. Then ϕ is a quotient map if the open sets of Y are the direct images of the saturated open sets of X.

A continuous surjection is necessarily a quotient map when the codomain has discrete topology. In general, however, the codomain of a continuous surjection will have non-open subsets of which the inverse images are open and non-closed subsets of which the inverse images are closed. For example, take the identity function on a set with the topology of the domain a strict refinement of the topology of the codomain.

Of course, Proposition 3.12 and other results about the coinduced topology apply to quotient maps as a special case. In addition, we have

Proposition 3.14

Let $\phi: X \to Y$ and $\psi: Y \to Z$ be continuous functions, where X, Y and Z are topological spaces. If $\psi\phi$ is a quotient map then so is ψ.

For let W be a subset of Z such that $\psi^{-1}W$ is open in Y. Then $\phi^{-1}\psi^{-1}W$ is open in X, since ϕ is continuous, and so W is open in Z, since $\psi\phi$ is a quotient map.

In particular, take $Z = X$ and $\psi\phi$ the identity. In that case ϕ is called a *right inverse* of ψ and Proposition 3.14 states that a continuous function is a quotient map if it admits a (continuous) right inverse. Examples will be given later.

Care must be exercised when the induced and coinduced topologies occur together. Thus let $\phi: X \to Y$ be a quotient map. If A is a subspace of X it is not in general true that the function $A \to \phi A$ determined by ϕ is a quotient map. Nor is it true that the product $X \times Z \to Y \times Z$ of ϕ and the identity on Z is a quotient map for all topological spaces Z. These problems will be further discussed, with examples, at a later stage.

3.4 Quotient spaces

Suppose now that we have a topological space X and an equivalence relation R on X. Consider the set X/R of equivalence classes together with the natural projection $\pi\colon X \to X/R$ which assigns each point of X to its equivalence class. The topology on X/R which makes π a quotient map is called the *quotient topology* and X/R, with this topology, is called a *quotient space*[7] of X.

For example, let X be a topological space and let A be a subspace of X. Impose an equivalence relation R on X so that all the points of A are related to each other but each of the points of $X - A$ is only related to itself. The resulting quotient space is known as the topological space obtained from X by collapsing A to a point.

For another example, let G be a discrete group of self-homeomorphisms of the topological space X. For each point x of X the subset $G \cdot x = \{gx\colon g \in G\}$ is called the *orbit* of x, under the action of G. If we identify points of X which lie in the same orbit the resulting quotient space is called the *orbit space* of X, under the action of G, and denoted by X/G. Note that a function $\phi\colon X \to Y$, for any Y, induces a function $\psi\colon X/G \to Y$ whenever ϕ is invariant with respect to the action. Moreover, an invariant continuous function ϕ induces a continuous function ψ, by Proposition 3.12.

As an illustration consider the n-sphere S^n with the group Z_2 consisting of the antipodal transformation and the identity. In this case, the quotient space S^n/Z_2 can be identified with real projective n-space. Complex and quaternionic projective spaces can be similarly represented as quotient spaces of spheres of appropriate dimensions.

An equivalence relation R on a set X determines, by restriction, an equivalence relation R' on each subset X' of X, so that X'/R' may be regarded as a subset of X/R. When X is a topological space it is not in general true that the relative topology which X'/R' obtains from the quotient topology of X/R agrees with the quotient topology of X'/R' itself. For example, take X to be the real line $\mathbb{R}$ and take X' to be the half-open interval $I' = [0, 1)$. Let us place two real numbers in the same equivalence class if they differ by an integer. Since the induced equivalence relation on I' is trivial we may identify the

---

[7] The terms identification topology and identification space are also in common use.

quotient set with I' itself. Then $[0, \frac{1}{2})$ is open in the quotient topology but not in the relative topology.

One of the main uses of the quotient topology occurs in the theory of topological groups. Thus let H be a subgroup of the topological group G. Consider the (left) factor set G/H together with the natural projection $\pi\colon G \to G/H$, which assigns each element to the coset to which it belongs. If we give G/H the quotient topology determined by π then G/H is called the (left) *factor space* of G by H.

It follows from Proposition 3.12 that for each element g of G the translation homeomorphism $g_\#\colon G \to G$ induces a homeomorphism $G/H \to G/H$, which we may refer to likewise as translation by g and denote by $g_\#$. Consequently, G/H is a homogeneous space.

The right factor space of G by H is defined similarly. Of course, the decomposition of G into left cosets is different, in general, from the decomposition of G into right cosets. However, the inversion homeomorphism $G \to G$ induces a homeomorphism between the left factor space and the right factor space.

Our next two results give necessary and sufficient conditions for the topology of the factor space to be (i) discrete or (ii) trivial.

Proposition 3.15

Let H be a subgroup of the topological group G. The factor space G/H is discrete if and only if H is open in G.

For if G/H is discrete then $[H]$ is open in G/H, hence H is open in G. Conversely, if H is open in G then each coset of H is open in G, hence each point of G/H is open and so G/H is discrete.

Proposition 3.16

Let H be a subgroup of the topological group G. The factor space G/H is trivial if and only if H is dense in G.

For suppose that H is dense in G. Then

$$G/H = \pi G = \pi \operatorname{Cl} H = \operatorname{Cl} \pi H = \operatorname{Cl} \pi(e),$$

from which the triviality of the topology follows at once. Conversely, suppose that H is not dense in G. Then there exists an element g of G and an open neighbourhood U of g which does not intersect H. Hence $U \cdot H$ does not intersect

H and so πU is an open neighbourhood of $\pi(g)$ which does not contain $\pi(e)$. The topology of G/H is therefore non-trivial.

As a further illustration of these ideas we prove

Proposition 3.17

Let G be a topological group. The factor space of the direct product $G \times G$ by the diagonal subgroup ΔG is homeomorphic to G.

For consider the division function $d: G \times G \to G$, given by $d(g_1, g_2) = g_1 \cdot g_2^{-1}$. This admits a right inverse $c: G \to G \times G$, where $c(g) = (g, e)$. Since both c and d are continuous it follows that d is a quotient map. Consequently, the continuous bijection $(G \times G)/\Delta G \to G$ induced by d is also a quotient map and therefore a homeomorphism.

Exercises

3.1. Let X be a topological space in which every finite subspace has the trivial topology. Show that X itself has the trivial topology. Is the corresponding assertion for the discrete topology true or false?

3.2. The topological space X is the union $A_1 \cup A_2 \cup \cdots$ of a countable family of subspaces A_n such that $A_n \subset \text{Int } A_{n+1}$ for each n. The function $\phi: X \to Y$, where Y is a topological space, is such that $\phi|A_n$ is continuous for each n. Show that ϕ is continuous.

3.3. Let H be a non-closed subgroup of the topological group G. Show that $\text{Cl } H - H$ is dense in $\text{Cl } H$.

3.4. Show that in the topological group G the closure $\text{Cl}\{e\}$ of the neutral element has the trivial topology.

3.5. Show that both the following functions are quotient maps
   (i) $\phi: \mathbb{R} \times \mathbb{R} \to \mathbb{R}$, where $\phi(x, y) = x + y^2$,
   (ii) $\phi: \mathbb{R} \times \mathbb{R} \to [0, \infty)$, where $\phi(x, y) = x^2 + y^2$.

   (Here $\mathbb{R}$ has the euclidean topology and $[0, \infty)$ the relative topology.)

3.6. Let X, Y be topological spaces. Let R be the equivalence relation on $X \times Y$ given by $(\xi, \eta)R(\xi', \eta')$ if and only if $\eta = \eta'$. Show that the quotient space $(X \times Y)/R$ is homeomorphic to Y.

3.7. On the unit interval $I = [0, 1] \subset \mathbb{R}$, with the euclidean topology, let R be the equivalence relation for which $I \cap \mathbb{Q}$ and its complement are the equivalence classes. Show that the quotient space I/R has the trivial topology.

3.8. Show that the real projective line S^1/Z_2 is homeomorphic to S^1.

3.9. Show that the function $\phi \colon S^2 \to \mathbb{R}^4$ given by

$$\phi(x_0, x_1, x_2) = (x_0^2, x_0 x_1, x_1^2 + x_0 x_2, x_1 x_2)$$

induces an embedding of the real projective plane S^2/Z_2 in $\mathbb{R}^4$, with the euclidean topology.

4
Open Functions and Closed Functions

4.1 General remarks

In Chapter 2 we studied functions from one topological space into another which are structure-preserving in the inverse image sense. In the present chapter we shall primarily be concerned with functions which are structure-preserving in the direct image sense. Specifically, we shall be concerned with functions which send open sets to open sets, and with functions which send closed sets to closed sets. Usually, but not invariably, the functions will be required to be continuous as well.

4.2 Open functions[8]

To begin with an example, let us return to the squaring function $\phi \colon \mathbb{R} \to \mathbb{R}$, given by

$$\phi(t) = t^2 \qquad (t \in \mathbb{R}),$$

where $\mathbb{R}$ denotes the real line with the euclidean topology. Of course ϕ is continuous, as we have seen, but the direct image $\phi\mathbb{R} = [0, +\infty)$ is not open although the full set $\mathbb{R}$ itself is open. Thus ϕ is not open, according to

---

[8] Also known as interior functions.

Definition 4.1

Let X and Y be topological spaces. The function $\phi: X \to Y$ is open if the direct image ϕU is open in Y whenever U is open in X.

For example, every function $\phi: X \to Y$ is open when Y is discrete, while every surjection $\phi: X \to Y$ is open when X is trivial.

Clearly, the identity function on a topological space is always open. Also if $\phi: X \to Y$ and $\psi: Y \to Z$ are open, where X, Y and Z are topological spaces, then the composition $\psi\phi: X \to Z$ is open.

Proposition 4.2

Let $\phi: X \to Y$ be a function, where X and Y are topological spaces. If the restriction $\phi | X_j$ is open for each member X_j of a covering of X then ϕ is open.

For if U is open in X then $U \cap X_j$ is open in X_j, by the relative topology, and so if $\phi(U \cap X_j)$ is open in Y, for each j, then $\phi U = \bigcup_j \phi(U \cap X_j)$ is open in Y.

Note that an open continuous injection is necessarily an embedding, and that an open continuous surjection is necessarily a quotient map. In the other direction we have

Proposition 4.3

Let X and Y be topological spaces.

(i)   *The embedding $\phi: X \to Y$ is open if (and only if) ϕX is open in Y.*
(ii) *The quotient map $\phi: X \to Y$ is open if (and only if) the saturation $\phi^{-1}\phi U$ is open in X whenever U is open in X.*

The proofs are obvious. Next we establish

Proposition 4.4

Let $\phi: X \to Y$ and $\psi: Y \to Z$ be functions, where X, Y and Z are topological spaces. Suppose that the composition $\psi\phi: X \to Z$ is open.

(i)   *If ψ is a continuous injection then ϕ is open.*
(ii) *If ϕ is a continuous surjection then ψ is open.*

In the case of (i), if U is open in X then $\psi\phi U$ is open in Z, hence $\phi U = \psi^{-1}\psi\phi U$ is open in Y.

In the case of (ii), if V is open in Y then $\phi^{-1}V$ is open in X, hence $\psi V = \psi\phi\phi^{-1}V$ is open in Z.

Proposition 4.5

Let X and Y be topological spaces. The function $\phi\colon X \to Y$ is open if and only if $\phi \operatorname{Int} H \subset \operatorname{Int} \phi H$ for each subset H of X.

For suppose that ϕ is open. If $H \subset X$ then $\operatorname{Int} H \subset H$ and so $\phi \operatorname{Int} H \subset \phi H$. However $\phi \operatorname{Int} H$ is open, since $\operatorname{Int} H$ is open, and so $\phi \operatorname{Int} H \subset \operatorname{Int} \phi H$, by definition of the interior.

Conversely, suppose that $\phi \operatorname{Int} H \subset \operatorname{Int} \phi H$ for all H. Take H to be open in X. Then $\phi H \subset \operatorname{Int} \phi H$ and so ϕH is open in Y.

Proposition 4.6

Let X and Y be topological spaces. If the function $\phi\colon X \to Y$ is open then for each subset M of Y and each closed set H of X containing $\phi^{-1}M$ there exists a closed set K of Y containing M such that $\phi^{-1}K \subset H$.

For take $K = Y - \phi(X - H)$. Then

$$\phi^{-1}K = X - \phi^{-1}[\phi(X - H)] \subset X - [X - H] = H.$$

If ϕ is open then $\phi(X - H)$ is open and so K is closed in Y. Also $M \subset K$, since $\phi^{-1}M \subset H$, which proves the result.

As we have remarked earlier, the inverse image behaves well in relation to the operations of union and intersection, while the direct image behaves well in relation to the operation of union but not the operation of intersection. This implies, in the topological situation, that a function may transform every member of a generating family into an open set and yet fail to be an open function. However, Proposition 4.2 at once implies

Proposition 4.7

Let $\phi\colon X \to Y$ be a function, where X and Y are topological spaces. Then ϕ is open if (and only if) ϕU is open in Y for each member U of a complete generating family for the topology of X.

For example, consider the addition function $\phi\colon \mathbb{R} \times \mathbb{R} \to \mathbb{R}$. Open intervals of the form (α, β) constitute a complete generating family for $\mathbb{R}$. Hence open slabs of the form $(\alpha_1, \beta_1) \times (\alpha_2, \beta_2)$ constitute a complete generating family for $\mathbb{R} \times \mathbb{R}$.

Now the direct image of such a slab with respect to ϕ is the open interval (α, β), where $\alpha = \alpha_1 + \alpha_2$, $\beta = \beta_1 + \beta_2$. Therefore the addition function is open, similarly the multiplication function is open.

Two important applications of Proposition 4.7 concern the topological product.

Corollary 4.8

Let $\{X_j\}$ be a family of topological spaces. Then the projection

$$\pi_j \colon \prod X_j \to X_j$$

is open, for each index j.

Corollary 4.9

Let $\{\phi_j\}$ be a family of open functions $\phi_j \colon X_j \to Y_j$, where X_j and Y_j are topological spaces. Then the product

$$\prod \phi_j \colon \prod X_j \to \prod Y_j$$

is open.

To illustrate these ideas we return to the theory of topological groups. Let H be a subgroup of the topological group G. Recall that the factor space G/H has the quotient topology determined by the natural projection $\pi \colon G \to G/H$. Moreover, if U is open in G then, as we have already seen, the saturation $\pi^{-1}\pi U = U \cdot H$ is open in G and so πU is open in G/H. Thus π is open, as well as continuous.

Now suppose that H is normal in G so that G/H has the factor group structure at the algebraic level. The inversion u' and multiplication m' in G/H are induced by the inversion u and multiplication m in G as indicated in the following diagram.

$$
\begin{array}{ccc}
G & \xrightarrow{\ u\ } & G \\
\pi \downarrow & & \downarrow \pi \\
G/H & \xrightarrow{\ u'\ } & G/H
\end{array}
\qquad
\begin{array}{ccc}
G \times G & \xrightarrow{\ m\ } & G \\
\pi \times \pi \downarrow & & \downarrow \pi \\
G/H \times G/H & \xrightarrow{\ m'\ } & G/H
\end{array}
$$

Now π is a quotient map, and so u' is continuous. In general, as stated above, the product of quotient maps is not a quotient map. In this case, however, π is open and so $\pi \times \pi$ is open. Therefore $\pi \times \pi$ is a quotient map and so m' is continuous. Thus we conclude that G/H is a group in the topological sense, with $\pi \colon G \to G/H$ a (continuous) homomorphism.

Now suppose that we have a (continuous) epimorphism $\phi \colon G \to G'$, where G and G' are topological groups. The kernel H of ϕ is a normal subgroup, of course, so that ϕ induces an isomorphism

$$\psi \colon G/H \to G',$$

in the algebraic sense. If ϕ is open then so is ψ, hence ψ is a homeomorphism and so an isomorphism in the topological sense.

For example, take $G = \mathbb{R}$ and take G' to be the circle group, consisting of complex numbers of unit modulus. The exponential function $t \mapsto e^{2\pi i t}$ is a homomorphism with kernel the subgroup of integers $\mathbb{Z}$. The function is a continuous open homomorphism, hence the factor group $\mathbb{R}/\mathbb{Z}$ is isomorphic to the circle group.

For another example, consider the determinant homomorphism

$$\det = D \colon \mathrm{Gl}(n, \mathbb{R}) \to \mathbb{R}_*.$$

Let A be an element of $\mathrm{Gl}(n, \mathbb{R})$ and let U be an open neighbourhood of A. Since U is open in $\mathrm{Gl}(n, \mathbb{R})$ there is an open interval J containing unity such that $tA \in U$ for all $t \in J$. As t ranges over J, t^n takes every value in some open interval J' containing unity, and so $t^n D(A) = D(tA)$ takes every value in some open interval J'' containing $D(A)$. Thus $D(U) \supset J''$ and $D(U)$ is open in $\mathbb{R}_*$. So D is open, as well as continuous. In this case the kernel $\mathrm{Sl}(n, \mathbb{R})$ is called the *modular group* and we see by the above that the factor group $\mathrm{Gl}(n, \mathbb{R})/\mathrm{Sl}(n, \mathbb{R})$ is isomorphic to $\mathbb{R}_*$, as a topological group.

4.3 Closed functions

We now turn from open functions to closed functions. Up to a point the two theories appear to be quite similar but there is a crucial difference, as we shall see, and it is this which gives rise to the idea of compactness.

Definition 4.10

Let X and Y be topological spaces. The function $\phi \colon X \to Y$ is closed if the direct image ϕE is closed in Y whenever E is closed in X.

For example, every function $\phi\colon X \to Y$ is closed when Y is discrete, while every surjection $\phi\colon X \to Y$ is closed when X is trivial.

Clearly the identity function on a topological space is always closed. Also if $\phi\colon X \to Y$ and $\psi\colon Y \to Z$ are closed, where X, Y and Z are topological spaces, then the composition $\psi\phi\colon X \to Z$ is closed.

Proposition 4.11

Let $\phi\colon X \to Y$ be a function, where X and Y are topological spaces. If the restriction $\phi|X_j$ is closed for each member X_j of a finite covering of X then ϕ is closed.

For if E is closed in X then $E \cap X_j$ is closed in X_j, by the relative topology, and so if $\phi(E \cap X_j)$ is closed in Y, for each j, then $\phi E = \bigcup_j \phi(E \cap X_j)$ is closed in Y, since the union is finite.

This shows, for example, that the squaring function $\phi\colon \mathbb{R} \to \mathbb{R}$ is closed, since the restrictions of ϕ to $(-\infty, 0]$ and $[0, \infty)$ are obviously closed.

Note that a closed continuous injection is necessarily an embedding, while a closed continuous surjection is necessarily a quotient map. In the other direction we have

Proposition 4.12

Let X and Y be topological spaces.

(i)   *The embedding $\phi\colon X \to Y$ is closed if (and only if) ϕX is closed in Y.*
(ii) *The quotient map $\phi\colon X \to Y$ is closed if (and only if) the saturation $\phi^{-1}\phi E$ is closed in X whenever E is closed in X.*

The proofs are obvious. Next we establish

Proposition 4.13

Let $\phi\colon X \to Y$ and $\psi\colon Y \to Z$ be functions, where X, Y and Z are topological spaces. Suppose that the composition $\psi\phi\colon X \to Z$ is closed.

(i)   *If ψ is a continuous injection then ϕ is closed.*
(ii) *If ϕ is a continuous surjection then ψ is closed.*

In the case of (i), if E is closed in X then $\psi\phi E$ is closed in Z, hence $\phi E = \psi^{-1}\psi\phi E$ is closed in Y.

In the case of (ii), if F is closed in Y then $\phi^{-1}F$ is closed in X, hence $\psi F = \psi\phi\phi^{-1}F$ is closed in Z.

Proposition 4.14

Let X and Y be topological spaces. The function $\phi\colon X \to Y$ is closed if and only if $\mathrm{Cl}\, \phi H \subset \phi\, \mathrm{Cl}\, H$ for each subset H of X.

For suppose that ϕ is closed. If $H \subset X$ then $H \subset \mathrm{Cl}\, H$ and so $\mathrm{Cl}\, \phi H \subset \mathrm{Cl}(\phi\, \mathrm{Cl}\, H) = \phi\, \mathrm{Cl}\, H$, since $\mathrm{Cl}\, H$ is closed.

Conversely, suppose that $\mathrm{Cl}\, \phi H \subset \phi\, \mathrm{Cl}\, H$ for all $H \subset X$. Take H to be closed in X. Then $\mathrm{Cl}\, \phi H \subset \phi H$ and so ϕH is closed.

Proposition 4.15

Let $\phi\colon X \to Y$ be a function where X and Y are topological spaces. If ϕ is closed then for each subset M of Y and open neighbourhood U of $\phi^{-1}M$ there exists an open neighbourhood V of M such that $\phi^{-1}V \subset U$. Conversely, suppose that for each point y of Y and open neighbourhood U of $\phi^{-1}(y)$ there exists an open neighbourhood V of y such that $\phi^{-1}V \subset U$; then ϕ is closed.

To prove the first part take $V = Y - \phi(X - U)$. Then

$$\phi^{-1}V = X - \phi^{-1}[\phi(X - U)] \subset X - [X - U] = U.$$

If ϕ is closed then $\phi(X - U)$ is closed and so V is open in Y. Also $M \subset V$ since $\phi^{-1}M \subset U$, as required. To prove the second part, let H be closed in X. If $y \notin \phi H$ then $X - H$ is an open neighbourhood of $\phi^{-1}(y)$. There exists, therefore, an open neighbourhood V of y such that $\phi^{-1}V$ does not meet H, hence V does not meet ϕH. So ϕH is closed.

Note that there is no analogue of Proposition 4.7, for closed functions, and so no analogues of Corollaries 4.8 and 4.9. In fact we have

Example 4.16

The projections

$$\pi\colon \mathbb{R} \times \mathbb{R} \to \mathbb{R}$$

are not closed, with the euclidean topology.

To see this consider, as in Figure 4.1, the closed set

$$H = \{(\xi, \eta) \in \mathbb{R} \times \mathbb{R}\colon \xi \cdot \eta = 1\}$$

of the domain; its image $\pi H = \mathbb{R}_*$ is not a closed set of the codomain.

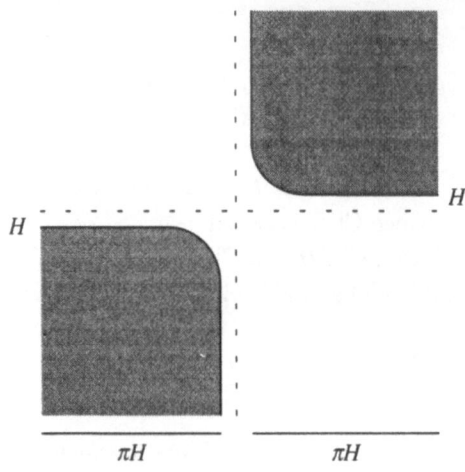

Figure 4.1.

Consideration of this significant example suggests that the failure of the projection to be closed may be related in some way to the fact that the domain is unbounded. To understand the situation better, let us extend the discussion to metric spaces generally.

Definition 4.17

A metric ρ on the set X is bounded if for some real number k we have

$$\rho(\xi, \eta) \le k \qquad (\xi, \eta \in X).$$

For example, the discrete metric is always bounded, with $k = 1$, while the euclidean metric on real n-space $(n \ge 1)$ is not. The reader may wish to consider which of the other examples of metrics mentioned above are bounded and which are not.

Definition 4.18

The subset H of the metric space X is bounded if for some real number k we have

$$\rho(\xi, \eta) \le k$$

for all $\xi, \eta \in H$.

In other words, H is bounded if and only if the induced metric on H is bounded. In that case a further definition suggests itself.

Definition 4.19

Let H be a bounded non-empty subset of the metric space X. The diameter of H is the real number

$$\text{diam } H = \sup \{\rho(\xi, \eta) \colon \xi, \eta \in H\}.$$

Clearly H is bounded if (and only if) the closure Cl H is bounded, and diam $H = \text{diam Cl } H$. It is not to be expected that points $\xi, \eta \in H$ can always be found such that $\rho(\xi, \eta) = \text{diam } H$. For example, take $X = \mathbb{R}$, with the euclidean metric, and take H to be the open interval $(0, 1)$, which has diameter 1 although $\rho(\xi, \eta) < 1$ for all $\xi, \eta \in (0, 1)$.

It is remarkable that any metric ρ on the set X can be transformed into a bounded metric ρ' without altering the metric topology, through formulae such as

$$\rho' = \min(1, \rho), \qquad \rho' = \rho(1 + \rho)^{-1}.$$

In the first case, for example, each open ε-ball ($\varepsilon > 0$) at a point x with respect to ρ contains the open ε'-ball at x with respect to ρ', where $\varepsilon' = \min(1, \varepsilon)$, while each ε'-ball ($\varepsilon' > 0$) with respect to ρ' contains an ε'-ball at x with respect to ρ. Consequently the neighbourhood filters are the same in the first case, and similarly in the second case.

Proposition 4.20

Suppose that the second projection $\pi \colon X \times \mathbb{R} \to \mathbb{R}$ is closed, where X is a metric space. Then X is bounded.

For consider the subset H of $X \times \mathbb{R}$ consisting of pairs (x, t) such that $\rho(x, x_0) \cdot t \geq 1$, where $x_0 \in X$ is fixed. Suppose, to obtain a contradiction, that X is not bounded. Then 0 adheres to πH, since if $\varepsilon > 0$ there exists a point (x, t) of H for which $\rho(x, x_0) \geq \varepsilon^{-1}$. But 0 is not a point of πH, and so πH is not closed. Since H is closed we have our contradiction.

We now consider the bounded interval $I = [0, 1] \subset \mathbb{R}$ and prove

Proposition 4.21

The second projection $\pi \colon I \times T \to T$ is closed for all topological spaces T.

We use the criterion of Proposition 4.15. Thus let t be a point of T and let U be an open neighbourhood of $I \times \{t\}$ in $I \times T$. We shall show that there exists an open neighbourhood W of t in T such that $I \times W \subset U$.

Let us say that a subset S of I is *bad* if there exists no open neighbourhood W of t for which $S \times W \subset U$. Note that if $S = S' \cup S''$ and S is bad then at least one of S', S'' must be bad. For if neither is bad then $S' \times W' \subset U$ and $S'' \times W'' \subset U$, for some open neighbourhoods W', W'', and then $S \times W \subset U$, where $W = W' \cap W''$.

Suppose, to obtain a contradiction, that I is bad. Write $I = I_0$. By the above remark, at least one of the intervals $[0, \frac{1}{2}]$, $[\frac{1}{2}, 1]$ is bad. Choose one and denote it by I_1. Subdividing I_1 in a similar fashion, choose I_2 in the same way, and so on. We obtain a nested chain

$$I = I_0 \supset I_1 \supset \cdots$$

of closed intervals, each of which is bad. The right-hand end-points $\xi_0, \xi_1, \ldots$ of these intervals form a monotone decreasing sequence of real numbers, bounded below by zero. The sequence converges to a limit ξ, say, where $\xi \geq 0$.

Now not every open neighbourhood of ξ can be bad, since by the product topology there exists an open neighbourhood V of ξ and an open neighbourhood W of t such that $V \times W \subset U$. But then V contains a bad interval, and so we have our contradiction.

Exercises

4.1. Let $\phi: X \to Y$ be a function, where X and Y are topological spaces. Let $\{Y_j\}$ be a family of subsets of Y which is either an open covering or a finite closed covering. Show that if the function $\phi^{-1}Y_j \to Y_j$ determined by ϕ is open (resp. closed) for each index j then ϕ is open (resp. closed).

4.2. Let $\phi: X \to Y$ be an open and closed function where X and Y are topological spaces. Let $\alpha: X \to I$ be continuous, where $I = [0, 1] \subset \mathbb{R}$, with the euclidean topology. Prove that $\beta: Y \to I$ is continuous, where

$$\beta(y) = \sup \{\alpha(x): x \in \phi^{-1}(y)\} \qquad (y \in Y).$$

4.3. By considering an appropriate subset of the domain, such as

$$\{(\xi, \eta): (\xi - \eta) \cdot (\xi + \eta) = 1\},$$

show that the function $\mathbb{R} \times \mathbb{R} \to \mathbb{R}$ given by real addition is not closed, with the euclidean topology.

4.4. By considering an appropriate subset of the domain, such as

$$\{(\xi, \eta) \colon \xi \cdot \eta \cdot (\xi + \eta) = 1\},$$

show that the function $\mathbb{R} \times \mathbb{R} \to \mathbb{R}$ given by real multiplication is not closed, with the euclidean topology.

4.5. Let H, K be bounded subsets of the metric space X such that H intersects K. Show the $H \cup K$ is also bounded and that

$$\operatorname{diam} (H \cup K) \leq \operatorname{diam} H + \operatorname{diam} K.$$

4.6. Show that if A is open (resp. closed) in X then the projection $\pi \colon X \to X/A$ is open (resp. closed).

5
Compact Spaces[9]

5.1 Introduction

Of all the invariants of topology compactness is undoubtedly the most important. To appreciate it fully one has to look at it in several different ways. The way I have chosen to define it in the first instance follows on very naturally from the discussion of closed functions at the end of the previous chapter. Once we have established its main properties and considered some examples we will show how compactness can be characterized in other ways. Of these the characterization in terms of the existence of finite subcoverings of open coverings is certainly the best known.

Definition 5.1

The topological space X is compact if the second projection $\pi \colon X \times T \to T$ is closed for all topological spaces T.

Thus I is compact, by Proposition 4.21, while $\mathbb{R}$ is not compact, by Example 4.16. Since compactness is obviously a topological invariant we conclude that $\mathbb{R}$ is not homeomorphic to I; more generally, that the open interval (α, β), where $\alpha < \beta$, is not homeomorphic to the closed interval $[\alpha, \beta]$. This is just the first of many applications of compactness.

---

[9] Bourbaki [2], amongst others, reserves the term compact to mean what we call compact Hausdorff, and uses the term quasicompact to mean what we call compact.

One-point spaces, more generally finite spaces, are compact, whatever the topology. Infinite discrete spaces are not compact, as we shall see in a moment. On the other hand, we have

Proposition 5.2

Let X be a cofinite space. Then X is compact.

For consider the projection $\pi: X \times T \to T$, where T is any topological space. Let t be a point of T and let U be an open neighbourhood of $X \times \{t\}$ in $X \times T$. By the product topology there exists, for each point x of X, an open neighbourhood V_x of x in X and an open neighbourhood W_x of t in T such that $V_x \times W_x \subset U$. Choose any x; then V_x is the complement of a finite set $\{x_1, \ldots, x_n\}$, say. The union

$$V = V_x \cup V_{x_1} \cup \cdots \cup V_{x_n}$$

contains both V_x and the complement of V_x and so coincides with X, while the intersection

$$W = W_x \cap X_{x_1} \cap \cdots \cap W_{x_n}$$

is an open neighbourhood of t in T. Thus $X \times W = V \times W \subset U$ and so π is closed, by Proposition 4.15.

When X is infinite there is a difference between the cofinite topology and the discrete topology; in fact, the latter is not compact. To see this let X be an infinite discrete space and consider the projection $\pi: X \times X^+ \to X^+$, where X^+ is the cofinite space consisting of the set X together with an additional point. In the topological product $X \times X^+$ consider the "diagonal" subset D consisting of pairs (x, x), where $x \in X$. Now $\pi D = X$, which is infinite and so not closed in X^+. But D itself is closed since if $x' \neq x$, where $x \in X$ and $x' \in X^+$, then $\{x\} \times (X^+ - \{x\})$ is an open neighbourhood of (x, x') which does not intersect D. Thus X is not compact as asserted.

The definition makes it clear that compactness is a topological invariant. Moreover, we have

Proposition 5.3

Let $\phi: X \to X'$ be a continuous surjection, where X and X' are topological spaces. If X is compact then so is X'.

For $\pi = \pi'(\phi \times \mathrm{id})$, as shown below, where π' denotes the projection in the case of X'.

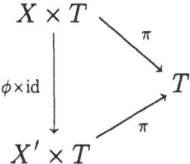

Now $\phi \times$ id is a continuous surjection, since ϕ is a continuous surjection. Hence π' is closed whenever π is closed, by Proposition 4.13(ii). The result follows at once.

Proposition 5.4

Let $\phi\colon X' \to X$ be a closed embedding, where X and X' are topological spaces. If X is compact then so is X'.

For $\phi \times$ id$\colon X' \times T \to X \times T$ is a closed embedding, since ϕ is a closed embedding, for each topological space T. Therefore $\pi' = \pi(\phi \times$ id$)$ is closed whenever π is closed, which proves the assertion.

We conclude that closed subspaces of compact spaces are compact. As we shall see in a moment the conclusion is untrue for subspaces in general. Note that Proposition 4.11 implies

Proposition 5.5

Let $\{X_j\}$ be a finite covering of the topological space X. If each of the X_j is compact then so is X.

Proposition 5.6 (Tychonoff)

Let $X_1, \ldots, X_n$ be compact spaces. Then the topological product $X_1 \times \cdots \times X_n$ is compact.

For let T be any topological space. Since X_n is compact the projection

$$X_n \times T \to T$$

is closed. Since X_{n-1} is compact the projection

$$X_{n-1} \times X_n \times T \to X_n \times T$$

is closed. So we continue until at last

$$X_1 \times X_2 \times \cdots \times X_n \times T \to X_2 \times \cdots \times X_n \times T$$

is closed, since X_1 is compact. Now compose all these closed maps, in the obvious way; we obtain that the projection

$$X_1 \times X_2 \times \cdots \times X_n \times T \to T$$

is closed. Therefore $X_1 \times \cdots \times X_n$ is compact, as asserted.

In fact, the topological product of an infinite family of compact spaces is compact, as we shall see later in this chapter.

We now come to the theorem of Heine–Borel, which characterizes compact subsets of the real n-space.

Theorem 5.7

Let X be a subspace of the real n-space $\mathbb{R}^n$, with euclidean topology. Then X is compact if and only if X is closed and bounded.

This result enables us to see at once that the closed n-ball B^n and the $(n-1)$-sphere S^{n-1} are compact, while the open n-ball U^n and the punctured $(n-1)$-sphere $S^{n-1} - \{p\}$ are not. A less obvious example will be discussed after we have given the proof of Theorem 5.7.

Consider first of all the composition

$$X \xrightarrow{\Gamma} X \times \mathbb{R}^n \xrightarrow{\pi} \mathbb{R}^n,$$

where Γ is the graph of the inclusion $\sigma \colon X \to \mathbb{R}^n$. Now ΓX is the inverse image of zero with respect to the continuous function

$$X \times \mathbb{R}^n \to \mathbb{R}^n,$$

given by vector subtraction, and is therefore closed. Thus Γ is always a closed embedding. When π is closed, as is the case for compact X, then $\sigma = \pi\Gamma$ is closed, i.e. X is closed in $\mathbb{R}^n$. Taken together with Proposition 4.20 this proves Theorem 5.7 in one direction.

For the proof in the other direction suppose that X is closed and bounded. Then X is contained in the closed slab

$$K = [\alpha_1, \beta_1] \times \cdots \times [\alpha_n, \beta_n],$$

where $\alpha_1 < \beta_1, \ldots, \alpha_n < \beta_n$. Now $[\alpha_j, \beta_j]$ $(j = 1, \ldots, n)$ is the continuous image of the unit interval I, under an affine transformation, and so is compact, by Propositions 4.21 and 5.3. Hence the topological product K is compact, by Proposition 5.6. Now X is closed in $\mathbb{R}^n$, by hypothesis, therefore closed in K, by the relative topology, and so compact, by Proposition 5.4. This completes the proof.

As a first application we deduce

Proposition 5.8

Let $\phi\colon X \to \mathbb{R}$ be continuous, where X is compact. Then ϕX is bounded, so that its infimum and supremum are defined. Moreover, there exist points ξ, η of X such that

$$\phi(\xi) = \sup(\phi X), \qquad \phi(\eta) = \inf(\phi X).$$

It is the last assertion which is meant by the expression: a continuous real-valued function on a compact space attains its bounds. The proof of Proposition 5.8 follows almost at once from the observation that ϕX is compact, therefore closed and bounded in $\mathbb{R}$. In fact, the only other remark necessary is that, as we have seen in Chapter 1, the distance between ϕX and both $\inf(\phi X)$ and $\sup(\phi X)$ is zero, since ϕX is closed.

For another illustration of the Heine–Borel theorem consider the group $\mathrm{O}(n, \mathbb{R})$ of real orthogonal $n \times n$ matrices. As with $\mathrm{Gl}(n, \mathbb{R})$ which contains $\mathrm{O}(n, \mathbb{R})$ as a subgroup, we topologize $\mathrm{O}(n, \mathbb{R})$ as a subspace of $\mathrm{M}(n, \mathbb{R})$, which is homeomorphic to $\mathbb{R}^m$ for $m = n \times n$. Now $\mathrm{O}(n, \mathbb{R})$ is bounded, as a subset of $\mathbb{R}^m$, since $x_{ij}^2 \le 1$ for each entry x_{ij} of an orthogonal matrix. Moreover, the orthogonality relations

$$\sum x_{jk} x_{jl} - \delta_{kl} = 0 \qquad (k, l = 1, 2, \ldots, n)$$

determine $\mathrm{O}(n, \mathbb{R})$ as the intersection of a number of closed sets of $\mathbb{R}^m$ so that $\mathrm{O}(n, \mathbb{R})$ is closed, as well as bounded. By Theorem 5.7, therefore, we conclude that $\mathrm{O}(n, \mathbb{R})$ is compact. On the other hand, $\mathrm{Gl}(n, \mathbb{R})$ is not compact, since $\mathbb{R}_*$ is the continuous image of $\mathrm{Gl}(n, \mathbb{R})$ under the determinant function and $\mathbb{R}_*$ is not compact.

Although Theorem 5.7 gives us a good idea of what compactness means for subsets of real n-space there is nothing comparable for metric spaces generally. We shall, however, be proving a result which gives at least some insight into the situation but first we must develop the theory further.

5.2 Compactness via filters

We now come to an important result which can be used to provide an alternative definition of compactness.

Theorem 5.9

The topological space X is compact if and only if each filter on X admits an adherence point.

For suppose that each filter on X admits an adherence point. Consider the projection

$$\pi \colon X \times T \to T,$$

where T is any topological space. Let E be a closed set of $X \times T$ and let t be an adherence point of πE. Then πE intersects every member of the neighbourhood filter $\mathcal{N}_t$ of t, and so E intersects every member of the induced filter $\pi^* \mathcal{N}_t$ on $X \times T$. Consider the trace $\mathcal{G}$ of $\pi^* \mathcal{N}_t$ on E. By hypothesis the direct image filter $p_* \mathcal{G}$ on X admits an adherence point $x \in X$, say, where $p \colon E \to X$ is given by the first projection. I assert that $(x, t) \in E$ and so $t \in \pi E$.

For suppose, to obtain a contradiction, that $(x, t) \notin E$. Since E is closed there exists a neighbourhood U of x in X and a neighbourhood V of t in T such that $U \times V$ does not intersect E. Now $E \cap (X \times V)$ is a member of $\mathcal{G}$, since V is a member of $\mathcal{N}_t$, and so $p(E \cap (X \times V))$ is a member of $p_* \mathcal{G}$. Also U is a neighbourhood of the adherence point x of $p_* \mathcal{G}$ and so U intersects $p(E \cap (X \times V))$. Therefore $U \times T$ intersects $E \cap (X \times V)$, in other words $U \times V$ intersects E, giving us our contradiction.

We conclude, therefore, that πE contains each of its adherence points and so is closed. This proves Theorem 5.9 in one direction.

Conversely, let $\mathcal{F}$ be a filter on X. Consider the set X' obtained by adjoining a point $*$ to X. Let $\mathcal{F}'$ be the filter on X' consisting of the sets $M \cup \{*\}$, where M runs through the members of $\mathcal{F}$. A coherent collection of filters on X' is formed by $\mathcal{F}'$, in the case of $*$, and by the principal filters ε'_x, for all points $x \in X$. Note that $*$ is an adherence point of X, in the resulting topology.

Now consider the graph D of the inclusion of X in X' as a subset of $X \times X'$ and write $\mathrm{Cl}\, D = E$. Since $\pi \colon X \times X' \to X'$ is closed, by assumption, we have $\pi E = \mathrm{Cl}\, \pi D = \mathrm{Cl}\, X$. Thus $*$ belongs to πE and so $(x, *)$ belongs to E for some point x of X. I assert that x is an adherence point of $\mathcal{F}$.

For if U is a neighbourhood of x in X and M is a member of $\mathcal{F}$ then $U \times (M \cup \{*\})$ is a neighbourhood of $(x, *)$ in $X \times X'$. Since $E = \mathrm{Cl}\, D$ and $(x, *) \in E$ this neighbourhood intersects D. Therefore U intersects $M \cup \{*\}$ and so intersects M. This proves the assertion and so completes the proof of Theorem 5.9.

Corollary 5.10

Let $\mathcal{F}$ be a filter on the compact space X, and let A be the set of adherence points of $\mathcal{F}$. Then each neighbourhood of A is a member of $\mathcal{F}$.

For let V be a neighbourhood of the adherence set A. Suppose, to obtain a contradiction, that each member of $\mathcal{F}$ intersects $X - V$. Then the trace of $\mathcal{F}$ on $X - V$ generates a filter $\mathcal{G}$ on X. Since X is compact, $\mathcal{G}$ admits at least one adherence point y, say. Now y cannot belong to A since the neighbourhood V of A does not intersect certain of the members of $\mathcal{G}$ (those which belong to the trace of $\mathcal{F}$). However, $\mathcal{G}$ is a refinement of $\mathcal{F}$, and so y is an adherence point of $\mathcal{F}$. Therefore $y \in A$, and we have our contradiction. Hence and from Proposition 1.27 we obtain

Corollary 5.11

The topological space X is compact if and only if each ultrafilter on X is convergent.

By Corollary 5.10 each filter on a compact space contains every neighbourhood of its adherence set. When the adherence set consists of a single point this yields

Proposition 5.12

Let $\mathcal{F}$ be a filter on the compact space X. Then $\mathcal{F}$ is convergent if $\mathcal{F}$ admits precisely one adherence point.

We are now almost ready to give a proof of the Tychonoff theorem, in full generality: the proof for finite products, given earlier, does not generalize to the infinite case and so a different approach is necessary. This depends on the characterization of ultrafilters we obtained in Chapter 1.

Proposition 5.13

Let $\mathcal{B}$ be a filter base on the set X. If for each subset M of X either $M \in \mathcal{B}$ or $X - M \in \mathcal{B}$ then $\mathcal{B}$ is an ultrafilter on X.

For let $\mathcal{F}$ be any filter refining the filter generated by $\mathcal{B}$. Then $\mathcal{F}$ coincides with $\mathcal{B}$, since if $M \in \mathcal{F}$ then $X - M \notin \mathcal{F}$, so $X - M \notin \mathcal{B}$, so $M \in \mathcal{B}$. So $\mathcal{B}$ is an ultrafilter, by Proposition 1.17, and we deduce

Proposition 5.14

Let $\phi: X \to Y$ be a function, where X and Y are sets. If $\mathcal{F}$ is an ultrafilter on X then $\phi_* \mathcal{F}$ is an ultrafilter on Y.

For let N be a subset of Y. If $\phi^{-1}N$ is a member of $\mathscr{F}$ then $N \in \phi_*\mathscr{F}$ since $\phi\phi^{-1}N \subset N$. If not then $X - \phi^{-1}N = \phi^{-1}(Y - N)$ is a member of $\mathscr{F}$ and hence $Y - N \in \phi_*\mathscr{F}$. Now apply Proposition 5.13.

Theorem 5.15

Let $\{X_j\}$ be a family of compact spaces. Then the topological product $\prod X_j$ is compact.

For let $\mathscr{F}$ be an ultrafilter on $X = \prod X_j$. Then $\pi_{j*}\mathscr{F}$ is an ultrafilter on X_j, for each index j, by Proposition 5.14. Since X_j is compact there exists a limit point x_j of $\pi_{j*}\mathscr{F}$. Then $x = (x_j)$ is a limit point of $\mathscr{F}$ and so X is compact.

One of the main applications of the theory of infinite products is in relation to function spaces. If X and Y are sets then the cartesian product Y^X of X copies of Y with itself is logically identical with the set of functions $\phi \colon X \to Y$; such a function can be regarded as a point of the cartesian Xth power. When Y (but not necessarily X) is a topological space we may give Y^X the product topology and any subset Φ of Y^X the relative topology.

Informally, we take all the functions in Φ which are close to a given function ϕ at finitely many points of X and these form a neighbourhood of ϕ in Φ, as suggested in Figure 5.1. Thus we can topologize sets of functions $\phi \colon X \to Y$ and, since there is more than one way of doing so, we refer to this as the topology of pointwise convergence, or pointwise topology. The reason for the name is provided by the following result, which is an immediate consequence of the definition of the topology.

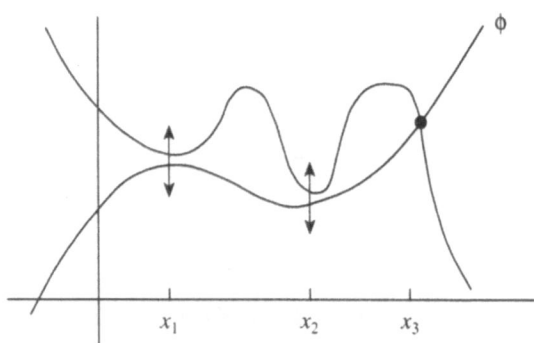

Figure 5.1.

Proposition 5.16

Let $\langle \phi_n \rangle$ be a sequence of functions in the topological space Φ. Then $\langle \phi_n \rangle$ converges to the function ϕ of Φ if and only if $\langle \phi_n(x) \rangle$ converges to $\phi(x)$ for each point x of X.

It is often important, in functional analysis, to know whether such a subset Φ of Y^X is compact. A useful pair of sufficient conditions are as follows:

(i) Φ is closed in Y^X,

(ii)   for each point x of X the subset

$$\{\phi(x) \colon \phi \in \Phi\}$$

has compact closure in Y.

In fact, the subset in (ii) is just the xth projection $\pi_x \Phi$ in Y, and we have

$$\prod_{x \in X} \mathrm{Cl}\, \pi_x \Phi = \mathrm{Cl} \prod_{x \in X} \pi_x \Phi.$$

If $\pi_x \Phi$ has compact closure, for each x, then so does $\prod_{x \in X} \pi_x \Phi$, by the Tyehnoff theorem. Now Φ is contained in $\prod_{x \in X} \pi_x \Phi$. If, further, Φ is closed in Y^X then Φ is a closed set of $\mathrm{Cl} \prod_{x \in X} \pi_x \Phi$, and so is compact.

5.3 Compactness via coverings

Our original definition of compactness was based on the notion of closed function. Then in Theorem 5.9 we characterized compact spaces through the existence of adherence points of filters. However, neither of these is the approach to compactness most commonly used, which is based on

Proposition 5.17

Let X be a topological space. Then X is compact if and only if each open covering of X admits a finite subcovering.

For suppose that X is compact. Let Γ be an open covering of X. Consider the dual family Γ^* of complements of members of Γ, which has empty intersection. I assert that some finite subfamily of Γ^* also has empty intersection. For suppose, to obtain a contradiction, that every finite subfamily has non-empty intersection. Then Γ^* generates a filter $\mathscr{F}$ on X. Now $\mathscr{F}$ admits an adherence point, by Theorem 5.9, since X is compact. The adherence point belongs to each of the closed members of $\mathscr{F}$; in particular, to each of the members of Γ^*. But the intersection of all the members of Γ^* is empty, since Γ covers X, and so we have our contradiction. If a finite subfamily of Γ^* has empty intersection then the corresponding subfamily of Γ covers X. This proves Proposition 5.17 in one direction. Examples are given in Figure 5.2, with $X = \mathbb{R}$, to show that this breaks down when X is not compact.

For the proof in the other direction suppose that each open covering of X admits a finite subcovering. Suppose, to obtain a contradiction, that there exists a filter $\mathscr{F}$ on X with no adherence point, i.e. the intersection of the family Γ^* of closures of members of $\mathscr{F}$ is empty. Then the dual family Φ of complements forms an open covering of X and so admits a finite subcovering, by hypothesis. Then the corresponding finite subfamily of Γ^* has empty intersection. But the members of Γ^* are all members of the filter $\mathscr{F}$, and so we have our contradiction. This completes the proof of Proposition 5.17.

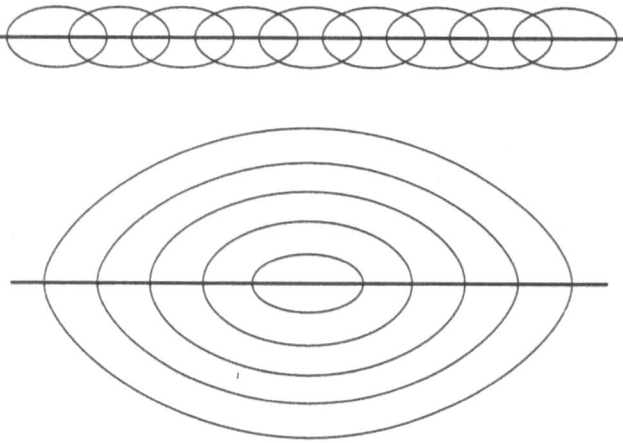

Figure 5.2. Two coverings of the real line.

The relationship between Theorem 5.9 and Proposition 5.17 is therefore quite close compared with that between Theorem 5.9 and our original Definition 5.1. It makes a good exercise to derive from Proposition 5.17 the results we proved earlier using Definition 5.1, such as the product theorem. It also makes a good exercise to take some of the results which seem easier to prove from Proposition 5.17, such as the following, and demonstrate them directly from Definition 5.1.

Proposition 5.18

Let Γ be an open covering of the compact metric space X. Then there exists a positive ε such that $U_\varepsilon(x)$ is contained in a member of Γ for each point x of X.

The number ε here is called the *Lebesgue number* of the covering; it is not unique, of course, since any positive number less than ε will have the same property.

To prove Proposition 5.18, first observe that each point x of X is contained in some member U of Γ and then, since U is open, there exists a positive number ε_x such that $U_{\varepsilon_x}(x) \subset U$. Consider the open covering $\{U_{\varepsilon'_x}(x) : x \in X\}$ of X, where $\varepsilon'_x = \frac{1}{2}\varepsilon_x$. Since X is compact we can extract a finite subcovering, indexed by $x_1, \ldots, x_n$, say. I assert that

$$\varepsilon = \min(\varepsilon'_{x_1}, \ldots, \varepsilon'_{x_n})$$

has the required property in relation to Γ. For each point x of X is contained in $U_{\varepsilon'_{x_j}}(x_j)$ for some j. So if $\rho(\xi, x) < \varepsilon$ then

$$\rho(\xi, x_i) \leq \rho(\xi, x) + \rho(x, x_j) < \varepsilon + \varepsilon'_{x_j} < \varepsilon_{x_j}.$$

So $U_\varepsilon(x) \subset U_{\varepsilon_{x_j}}(x_j) \subset U$, as required.

Exercises

5.1. Show that if X is a non-compact subset of the real line $\mathbb{R}$ then

   (i) there exists a continuous function $\phi \colon X \to \mathbb{R}$ which is not bounded,

   (ii) there exists a continuous function $\phi \colon X \to \mathbb{R}$ which is bounded but does not attain its bounds.

5.2. Let $\mathscr{F}$ be a family of closed compact subsets of the topological space X. Suppose that the intersection of the members of the family is contained in some open set U. Show that the intersection of the members of some finite subfamily is also contained in U.

5.3. Let E and F be compact subsets of the topological spaces X and Y, and let W be a neighbourhood of $E \times F$ in $X \times Y$. Show that there exist neighbourhoods U of E in X and V of F in Y such that $U \times V \subset W$.

5.4. Let X be a topological space. Let X^+ be the union of X and an additional point * with the topology consisting of (i) the open sets of X and (ii) the complements in X^+ of the closed compact sets of X. Show that X^+ is compact.

5.5. A continuous function $\phi\colon X \to Y$, where X and Y are topological spaces, is said to be proper if the product

$$\phi \times \mathrm{id}\colon X \times T \to Y \times T$$

is closed for all topological spaces T. Show that ϕ is proper if and only if ϕ is closed and $\phi^{-1}(y)$ is compact for all points y of Y.

5.6. Let A, B be compact subsets of the topological group G. Show that $A \cdot B$ is compact.

5.7. Let C be a compact subset and let E be a closed subset of the topological group G. Show that $C \cdot E$ is closed in G.

5.8. Let H be an open subgroup of the compact group G. Show that H is of finite index in G.

5.9. Let U be any open neighbourhood of the neutral element e in the topological group G, and let C be any compact subset of G. Show that there exists an open neighbourhood V of e for which $x \cdot V \cdot x^{-1} \subset U$ for all $x \in C$.

5.10. Let H be a compact subgroup of the topological group G. Show that the natural projection $\pi\colon G \to G/H$ is closed.

6

Separation Conditions

6.1 General remarks

We now come to a series of conditions, often regarded as axioms, all of which are satisfied by metric spaces with the metric topology. Traditionally these conditions tend to be labelled $T_0, T_1, \ldots$; however, the only case where we shall use this notation is for the condition known as T_1. Each condition is obviously a topological invariant.

6.2 T_1 spaces

We now state the first of the separation conditions to be considered here.

Definition 6.1

The topological space X is T_1 if $\{x\}$ is closed for each point x of X.

For example, discrete spaces and cofinite spaces are T_1 while trivial spaces with more than one point are not. Topological products of T_1 spaces are T_1. Subspaces of T_1 spaces are T_1. All these statements follow at once from the definition.

Let X be a topological space. For each point x of X the intersection of the neighbourhoods of x is the same as the intersection of the open neighbourhoods of x. The intersection need not coincide with the one-point set $\{x\}$; in the trivial topology, for example, the intersection is the full set X. In general the intersection of the open neighbourhoods is a proper subset of the intersection of the closed neighbourhoods.

Proposition 6.2

The topological space X is T_1 if and only if each point x of X coincides with the intersection of its (open) neighbourhoods.

For suppose that the intersection condition is satisfied. Given x, if $x' \neq x$ then x' has a neighbourhood which does not contain x, and so $\{x\}$ is closed.

In the other direction, suppose that for some point x the intersection of the neighbourhoods of x contains a point $x' \neq x$. Then $\{x'\}$ cannot be closed since that would imply the existence of a neighbourhood of x not containing x'.

6.3 Hausdorff spaces

The second condition is much more important than the first. Although the term T_2 is often used it is better known as the Hausdorff property.

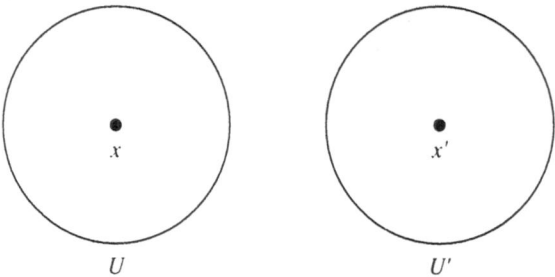

Figure 6.1. The Hausdorff condition.

Definition 6.3

The topological space X has the Hausdorff property if for each pair of distinct points x, x' of X there exist neighbourhoods U of x and U' of x' which are disjoint.

The situation is illustrated in Figure 6.1. Here and in similar definitions in this section it makes no real difference whether we write neighbourhood or open neighbourhood, or indeed basic neighbourhood where neighbourhood bases are given.

For example, discrete spaces are Hausdorff since we can take $U = \{x\}$, $U' = \{x'\}$. Infinite cofinite spaces, on the other hand, are never Hausdorff. Thus although Hausdorff spaces are obviously T_1 the converse is false.

For another example, let X be a metric space with metric ρ. Then $U_\varepsilon(x)$, $U_\varepsilon(x')$ are disjoint neighbourhoods of x, x', where $\varepsilon = \frac{1}{2}\rho(x, x')$.

Proposition 6.4

The topological space X is Hausdorff if and only if for each point x of X the intersection of the closed neighbourhoods of x is the one-point set $\{x\}$.

For let X be Hausdorff. Suppose, to obtain a contradiction, that for some point x there exists a distinct point x' such that each closed neighbourhood of x contains x'. By the Hausdorff property there exist disjoint neighbourhoods U of x and U' of x'. Then $X - U'$ is closed and contains the neighbourhood U of x, hence is a neighbourhood of x, contrary to hypothesis.

Conversely, suppose that the intersection of the closed neighbourhoods of x is just $\{x\}$ for each point x of X. Then if $x' \neq x$ there exists a closed neighbourhood N of x which does not contain x'. Hence N and $X - N$ are disjoint neighbourhoods of x and x', respectively. This completes the proof of Proposition 6.4.

Proposition 6.5

Let $\phi\colon X \to Y$ be a continuous injection, where X and Y are topological spaces. If Y is Hausdorff then so is X.

For let x, x' be distinct points of X. Then $\phi(x), \phi(x')$ are distinct points of Y, since ϕ is injective. Since Y is Hausdorff there exist disjoint neighbourhoods V, V' of $\phi(x), \phi(x')$, respectively. Then $\phi^{-1}V, \phi^{-1}V'$ are disjoint neighbourhoods of x, x', respectively.

We see from Proposition 6.5 that each subspace of a Hausdorff space is itself Hausdorff. We also see that the Hausdorff property is a topological invariant.

Proposition 6.6

Let $\{X_j\}$ be a family of Hausdorff spaces. Then the topological product $\prod X_j$ is a Hausdorff space.

For let $x = (x_j), x' = (x'_j)$ be distinct points of $X = \prod X_j$. Then $x_j \neq x'_j$ for some index j. Since X_j is Hausdorff there exist disjoint neighbourhoods U_j, U'_j

of x_j, x'_j in X_j. Their inverse images with respect to the jth projection $X \to X_j$ are disjoint neighbourhoods of x, x', as required.

Recall that the diagonal subset of $X \times X$ is denoted by ΔX. The complement of the diagonal consists of precisely the pairs (x, x') of distinct points of X. We prove

Proposition 6.7

The topological space X is Hausdorff if and only if ΔX is closed in $X \times X$.

For let (x, x') be a pair of distinct points of X. If ΔX is closed in $X \times X$ then (x, x') admits an open neighbourhood which does not intersect ΔX, hence a product open neighbourhood $U \times U'$ say with the same property. Then U, U' are disjoint open neighbourhoods of x, x' in X. This proves the result in one direction; the argument in the other direction is just the reverse.

Proposition 6.8

Let $\phi \colon X \to Y$ be continuous, where X and Y are topological spaces. If Y is Hausdorff then the graph embedding

$$\Gamma_\phi \colon X \to X \times Y$$

is closed.

For we have $\Gamma_\phi X = (\phi \times \mathrm{id})^{-1} \Delta Y$, as indicated, where ΔY is closed.

$$
\begin{array}{ccc}
X & \xrightarrow{\ \Gamma_\phi\ } & X \times Y \\
\downarrow{\scriptstyle \phi} & & \downarrow{\scriptstyle \phi \times \mathrm{id}} \\
Y & \xrightarrow{\ \Delta\ } & Y \times Y
\end{array}
$$

Proposition 6.9

Let ϕ, $\psi \colon X \to Y$ be continuous, where X and Y are topological spaces. If Y is Hausdorff then the coincidence set

$$M = \{x \in X \colon \phi(x) = \psi(x)\}$$

of ϕ and ψ is closed in X.

For we have $M = \Delta^{-1}(\phi \times \psi)^{-1} \Delta Y$, as indicated below.

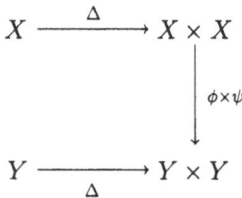

Now ΔY is closed, since Y is Hausdorff, and so M is closed, by continuity of $(\phi \times \psi)\Delta$.

Corollary 6.10

Let $\phi, \psi \colon X \to Y$ be continuous, where X and Y are topological spaces. If Y is Hausdorff and if ϕ and ψ agree on a dense subset of X then $\phi = \psi$.

Corollary 6.11

Let G be a Hausdorff group and let H be a commutative subgroup of G. Then the subgroup $\mathrm{Cl}\ H$ of G is also commutative.

To see this, write $\mathrm{Cl}\ H = \bar{H}$ and consider the continuous functions

$$m, m' \colon \bar{H} \times \bar{H} \to G,$$

where $m(x, x') = x \cdot x'$ and $m'(x, x') = x' \cdot x$. Since H is commutative the functions agree on $H \times H$, but $H \times H$ is dense in $\bar{H} \times \bar{H}$ and so $m = m'$, as required.

Our next result concerns the behaviour of filters on a Hausdorff space.

Proposition 6.12

The topological space X is Hausdorff if and only if the following condition holds for each convergent filter $\mathscr{F}$ on X: if x is a limit point of $\mathscr{F}$ then each adherence point of $\mathscr{F}$ coincides with x.

This shows, in particular, that limit points are unique in a Hausdorff space.

To prove Proposition 6.12, suppose that the stated condition is satisfied. Let ξ, η be distinct points of X. The neighbourhood filter $\mathscr{N}_\xi$ of ξ converges to ξ and so, by the condition, η cannot be an adherence point of $\mathscr{N}_\xi$. In other words, there exists a neighbourhood V of η and a neighbourhood U of ξ which are disjoint: thus X is Hausdorff.

In the other direction, suppose that there exists a filter $\mathscr{F}$ on X and distinct points ξ, η of X such that ξ is a limit point of $\mathscr{F}$ and η is an adherence point of $\mathscr{F}$. Then each neighbourhood V of η meets each neighbourhood U of ξ, and so X cannot be Hausdorff.

It is often important to know whether a quotient space of a topological space X is a Hausdorff space. The following result covers at least some of the situations which can arise in practice.

Proposition 6.13

Let R be an equivalence relation on the topological space X. The quotient space X/R is a Hausdorff space if the natural projection $\pi: X \to X/R$ is open and R is closed in $X \times X$.

For let x, x' be points of X such that $\pi(x) \neq \pi(x')$. Then x is not related to x', i.e. $(x, x') \notin R$. Since R is closed in $X \times X$ there exist neighbourhoods U, U' of x, x' in X such that $U \times U'$ does not intersect R. Then $\pi U, \pi U'$ are disjoint neighbourhoods of $\pi(x), \pi(x')$ in X/R, since π is open.

Proposition 6.14

Let H be a subgroup of the topological group G. Then the factor space G/H is Hausdorff if and only if H is closed in G.

For if G/H is Hausdorff (or even T_1) then $[H]$ is closed in G/H and so $H = \pi^{-1}[H]$ is closed in G. The converse follows from Proposition 6.13 but it makes a good exercise to deduce it from first principles. Note in particular that G itself is Hausdorff if and only if $\{e\}$ is closed in G.

We now come to a series of results involving both compactness and the Hausdorff property.

Proposition 6.15

Let $\phi: X \to Y$ be continuous, where X is compact and Y is Hausdorff. Then ϕ is closed.

For if Y is Hausdorff the graph embedding $X \to X \times Y$ is closed, as we have seen, and since X is compact the projection $X \times Y \to Y$ is closed; now Proposition 6.15 follows since the composition of closed functions is closed.

In particular, we see that compact subspaces of Hausdorff spaces are closed. Consequently, for compact Hausdorff spaces, a subset is closed if and only if it is compact.

A common procedure which depends on Proposition 6.15 for its justification is as follows. Let $\phi\colon X \to Y$ be a continuous surjection, where X and Y are topological spaces. In the equivalence relation $R = (\phi \times \phi)^{-1}\Delta Y$ on X determined by ϕ, points of X are equivalent if and only if they have the same image under ϕ, and so ϕ induces a continuous bijection $\psi\colon X/R \to Y$. If X is compact then X/R is compact and so if Y is also Hausdorff then ψ is a homeomorphism.

For example, consider the continuous surjection $\phi\colon S^n \times I \to B^{n+1}$, where $I = [0,1]$ and where

$$\phi(x,t) = tx \qquad (t \in I, x \in S^n).$$

The corresponding equivalence relation identifies all the points of $S^n \times \{0\}$. Now $S^n \times I$ is compact, since S^n and I are compact, and B^{n+1} is Hausdorff. We see, therefore, that B^{n+1} is homeomorphic to the quotient space of $S^n \times I$ which is obtained by collapsing $S^n \times \{0\}$.

Similarly, using a suitable function defined in terms of polar coordinates, we can represent S^{n+1} as a quotient space of the cylinder $S^n \times I$. In this case the equivalence relation identifies all points of $S^n \times \{0\}$ to one of the poles of S^{n+1} and all points of $S^n \times \{1\}$ to the other. The case $n = 2$ is illustrated in Figure 6.2.

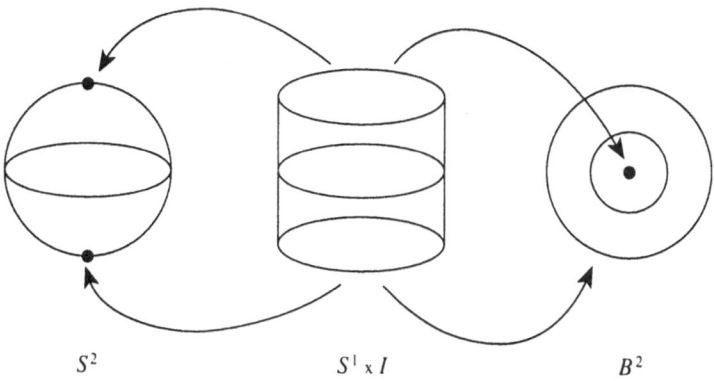

$S^2 \qquad\qquad\qquad S^1 \times I \qquad\qquad\qquad B^2$

Figure 6.2.

For another example, consider the topological group $O(n, \mathbb{R})$ of real orthogonal $n \times n$ matrices. A continuous surjection $\phi\colon O(n, \mathbb{R}) \to S^{n-1}$ can be defined by applying each matrix to a given point $p \in S^{n-1}$. In this case the equivalence classes are just the cosets of the subgroup $\phi^{-1}(p)$ of $O(n, \mathbb{R})$, which may be identified with $O(n-1, \mathbb{R})$. Since $O(n, \mathbb{R})$ is compact, as we have seen, and since S^{n-1} is Hausdorff we conclude that S^{n-1} can be represented as the factor space $O(n, \mathbb{R})/O(n-1, \mathbb{R})$.

6.4 Regular spaces[10]

We now move on to another of the separation conditions, which is neither weaker nor stronger than the Hausdorff property.

Definition 6.16

The topological space X is regular if for each point x of X and each neighbourhood U of x there exists a neighbourhood V of x such that Cl $V \subset U$.

In other words, the closed neighbourhoods of a point form a neighbourhood base at that point. Another way to express the condition, which is obviously equivalent, is as follows: for each point x of X and each closed set E such that $x \notin E$, there exists a neighbourhood V of x and a neighbourhood W of E which are disjoint. The situation is illustrated in Figure 6.3.

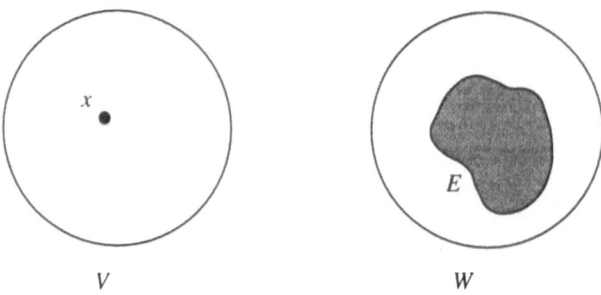

Figure 6.3. The regularity condition.

Discrete spaces are regular. Trivial spaces are also regular although not, in general, Hausdorff. Regular T_1 spaces are Hausdorff, of course; since infinite cofinite spaces are not Hausdorff they are not regular either. A refinement of a regular Hausdorff topology, although necessarily Hausdorff, may fail to be regular. For example, consider the real line $\mathbb{R}$ with the refinement of the euclidean topology in which the set $\mathbb{Q}$ of rationals is an additional open neighbourhood of unity. In the refined topology there exists no open neighbourhood V of unity such that Cl $V \subset \mathbb{Q}$.

Metric spaces are regular. For if X in Definition 6.16 has metric ρ then $U_\varepsilon(x)$ and $U_\varepsilon(E)$ are disjoint neighbourhoods of x and E, where $\varepsilon = \frac{1}{2}\rho(x, E)$.

---

[10] Some authors, including Bourbaki, require regular spaces to be Hausdorff as well.

Proposition 6.17

Let $\phi\colon X \to X'$ be an embedding, where X and X' are topological spaces. If X' is regular then so is X.

For let x be a point of X and let U be an open neighbourhood of x in X. Since ϕ is an embedding there exists an open neighbourhood U' of $\phi(x)$ in X' such that $\phi^{-1}U' = U$. If X' is regular there exists an open neighbourhood V' of $\phi(x)$ in X' such that $\operatorname{Cl} V' \subset U'$. Then $V = \phi^{-1}V'$ is an open neighbourhood of x in X and

$$\phi \operatorname{Cl} V \subset \operatorname{Cl} \phi V = \operatorname{Cl} \phi\phi^{-1}V' \subset \operatorname{Cl} V' \subset U';$$

thus $\operatorname{Cl} V \subset \phi^{-1}U' = U$, as required.

We see, in particular, that subspaces of regular spaces are regular. We also see that regularity is a topological invariant.

It should be observed that if the topology of X is generated by a complete family Γ then the condition in Definition 6.16 holds for all open neighbourhoods U if it holds for the Γ-neighbourhoods. We use this in the proof of

Proposition 6.18

Let $\{X_j\}$ be a family of regular spaces. Then the topological product $\prod X_j$ is regular.

For let $x = (x_j)$ be a point of the topological product and let $\prod U_j$ be a restricted product open neighbourhood of (x_j). Since each of the factors X_j is regular, and since U_j is an open neighbourhood of x_j in X_j, there exists an open neighbourhood V_j of x_j in X_j such that $\operatorname{Cl} V_j \subset U_j$; we take $V_j = X_j$ for those j such that $U_j = X_j$. Then $\prod V_j$ is a restricted product open neighbourhood of x such that $\operatorname{Cl} \prod V_j = \prod \operatorname{Cl} V_j \subset \prod U_j$, as required.

Topological groups and, more generally, factor spaces of topological groups, are necessarily regular. Specifically, we have

Proposition 6.19

Let H be a subgroup of the topological group G. Then the factor space G/H is regular.

Using homogeneity we can assume, without real loss of generality, that the point of G/H concerned is the neutral coset $[H]$. Let V be an open neighbourhood of $[H]$ in G/H. Then $U = \pi^{-1}V$ is an open neighbourhood of H, and hence of e, in G. Let N be an open neighbourhood of e such that

$N^{-1} \cdot N \subset U$. Then $\pi N = \pi(N \cdot H)$ is an open neighbourhood of $[H]$ in G/H. I assert that

$$\text{Cl}(\pi N) \subset \pi(N^{-1} \cdot N) \subset \pi U \subset V.$$

For if $[gH]$ is an adherence point of πH, where $g \in G$, the saturated neighbourhood $N \cdot gH$ of gH meets $N \cdot H$, hence gH meets $N^{-1} \cdot N \cdot H$, hence $[gH] \in \pi(N^{-1} \cdot N)$, as required. Thus G/H is regular, as asserted.

Proposition 6.20

Let X be a compact Hausdorff space. Then X is regular.

For let x be a point of X, and let $\mathscr{B}_x$ be the filterbase formed by the *closed* neighbourhoods of x. Since X is Hausdorff the intersection of the members of $\mathscr{B}_x$ is $\{x\}$. Therefore x is the sole adherence point of $\mathscr{B}_x$, and so $\mathscr{B}_x$ converges to x by Proposition 5.12. Now regularity follows at once.

Given a topological space X, let us say that the point ξ is related to the point η if every neighbourhood of ξ contains η. The relation thus defined is reflexive and transitive, but not in general symmetric. Suppose, however, that X is regular. If U is a neighbourhood of ξ which does not contain η then there exists a neighbourhood V of ξ such that $\text{Cl}\, V \subset U$ and hence a neighbourhood $X - \text{Cl}\, V$ of η which does not contain ξ. For regular X, therefore, the relation is symmetric, as well as reflexive and transitive, and so is an equivalence relation.

It is easy to see that both the open sets and the closed sets of X are saturated, with respect to the relation. For if ξ is related to a point of an open set U then U, as a neighbourhood of that point, must contain ξ. Similarly, if ξ is a point of the complement of a closed set E then ξ has a neighbourhood which is disjoint from E, and so ξ is not related to any point of E, by transitivity. Consequently, the natural projection $\pi \colon X \to X'$ to the quotient space is both open and closed. Moreover, X' is Hausdorff, since if ξ, η are points of X such that $\pi(\xi) \neq \pi(\eta)$ then $\xi \notin \text{Cl}\{\eta\}$ and so, by regularity, there exist disjoint neighbourhoods U of ξ and V of $\text{Cl}\{\eta\}$ in X, hence disjoint neighbourhoods πU of $\pi(\xi)$ and πV of $\pi(\eta)$ in X'. Similarly, X' is regular.

We refer to X' as the *Hausdorff quotient* of X. The construction has the following characteristic property:

Proposition 6.21

Let $\phi \colon X \to Y$ be continuous, where X is regular and Y is Hausdorff. Then there exists one and only one continuous function $\psi \colon X' \to Y$ such that $\psi \pi = \phi$.

Of course, π is a homeomorphism when X itself is Hausdorff. Also X' is a one-point space when X has the trivial topology.

For another example, consider the case of a topological group G, which is always regular as we have seen. The Hausdorff quotient of G is the factor group $G/\mathrm{Cl}\{e\}$. More generally, the Hausdorff quotient of the factor space G/H, where H is a subgroup of G, is the factor space $G/\mathrm{Cl}\,H$.

As we observed earlier, it is often important to know whether or not a quotient space has the Hausdorff property. In Proposition 6.13 we have obtained a result of this nature in the case of a closed equivalence relation. We now prove another result of the same type.

Proposition 6.22

Let A be a closed subspace of the regular Hausdorff space X. Then the quotient space X/A obtained by collapsing A to a point is a Hausdorff space.

For let ξ, η be distinct points of X/A. If neither one is the point $[A]$ the existence of disjoint open neighbourhoods follows at once from the Hausdorff property of $X - A$. If one of ξ, η is the point $[A]$, say $\eta = [A]$, then $\pi^{-1}(\xi)$ is a single point and $X - A$ is an open neighbourhood of that point. By regularity there exists an open neighbourhood V of $\pi^{-1}(\xi)$ in X such that $\mathrm{Cl}\,V \subset X - A$. Then V is saturated, since V is disjoint from A, and so πV is open in X/A. Also $X - \mathrm{Cl}\,V$ contains A and is therefore also saturated, so that $\pi(X - \mathrm{Cl}\,V)$ is open in X/A. Thus πV is an open neighbourhood of ξ and $\pi(X - \mathrm{Cl}\,V)$ is an open neighbourhood of η. Since these sets are disjoint this completes the proof that X/A is Hausdorff.

6.5 Normal spaces

We now reach the last of the separation conditions to be considered here.

Definition 6.23

The topological space X is normal if each pair E, F of disjoint closed sets of X admit disjoint neighbourhoods U, V.

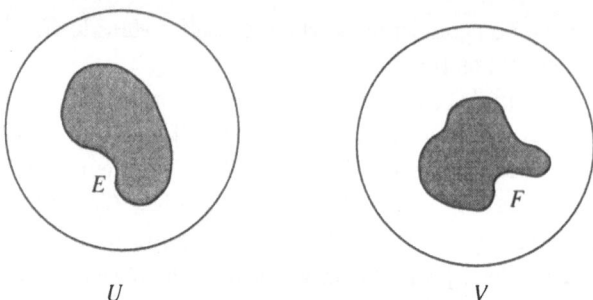

Figure 6.4. The normal condition.

The situation is illustrated in Figure 6.4. Clearly the condition is topologically invariant. Discrete spaces are normal: take $U = E$ and $V = F$. Trivial spaces are normal since the only possibilities for E and F are the empty set and the full set. Normal T_1 spaces are regular and Hausdorff; since infinite cofinite spaces are not Hausdorff they are not normal either. The Sierpinski point-pair is normal although it is neither regular nor Hausdorff. Metric spaces are normal. For if X in Definition 6.23 has metric ρ then U, V can be defined by

$$U = \{x \in X \colon \rho(x, E) < \rho(x, F)\}, \qquad V = \{x \in X \colon \rho(x, E) > \rho(x, F)\}.$$

Proposition 6.24

Let $\phi \colon X \to Y$ be a closed embedding, where X and Y are topological spaces. If Y is normal then so is X.

For let E, F be disjoint closed sets of X. Then ϕE, ϕF are disjoint closed sets of Y. Since Y is normal there exist disjoint neighbourhoods U, V of ϕE, ϕF in Y. Then $\phi^{-1}U$, $\phi^{-1}V$ are disjoint neighbourhoods of E, F in X.

We see, in particular, that closed subspaces of normal spaces are normal; examples can be given to show that subspaces in general are not necessarily normal.

Proposition 6.25

Let $\phi \colon X \to Y$ be a closed continuous surjection, where X and Y are topological spaces. If X is normal then so is Y.

For let E, F be disjoint closed sets of Y. Then $\phi^{-1}E$, $\phi^{-1}F$ are disjoint closed sets of X. Since X is normal there exist disjoint open neighbourhoods U, V of $\phi^{-1}E$, $\phi^{-1}F$ in X. Then $Y - \phi(X - U)$, $Y - \phi(X - V)$ are disjoint open neighbourhoods of E, F in Y.

Examples can be given to show that topological products of normal spaces are not necessarily normal (there is even an example where X is normal but the cylinder $X \times I$ is not).

Proposition 6.26

Let E, F be disjoint compact subsets of the Hausdorff space X. Then there exist open neighbourhoods U of E and V of F, which are disjoint.

For let ξ be any point of E. By the Hausdorff property there exist, for each point η of F, disjoint open neighbourhoods $U_\eta(\xi)$ of ξ, $V_\eta(\xi)$ of η. The sets $V_\eta(\xi)$ ($\eta \in F$) form an open covering of the compact F; extract a finite subcovering indexed by $\eta_1, \ldots, \eta_n$, say. Then the intersection

$$U(\xi) = U_{\eta_1}(\xi) \cap \cdots \cap U_{\eta_n}(\xi)$$

is an open neighbourhood of ξ which does not intersect the union

$$V(\xi) = V_{\eta_1}(\xi) \cup \cdots \cup V_{\eta_n}(\xi)$$

which is an open neighbourhood of F. Now vary ξ. The sets $U(\xi)$ ($\xi \in E$) form an open covering of the compact E; extract a finite subcovering indexed by $\xi_1, \ldots, \xi_m$, say. Then the union

$$U = U(\xi_1) \cup \cdots \cup U(\xi_m)$$

is an open neighbourhood of E which does not intersect the open neighbourhood

$$V = V(\xi_1) \cap \cdots \cap V(\xi_m)$$

of F.

Corollary 6.27

If X is a compact Hausdorff space then X is normal.

Proposition 6.28

Let C be a compact subset of the regular space X and let U be an open neighbourhood of C in X. Then $\mathrm{Cl}\, V \subset U$ for some open neighbourhood V of C in X.

Since U is an open neighbourhood of each point x of C there exist, by regularity, open neighbourhoods V_x of x and W_x of $X - U = E$ which are disjoint. The family $\{V_x : x \in C\}$ forms an open covering of the compact C; extract a finite subcovering indexed by $x_1, \ldots, x_m$, say. Then the union

$$V = V_{x_1} \cup \cdots \cup V_{x_m}$$

is an open neighbourhood of C and the intersection

$$W = W_{x_1} \cap \cdots \cap W_{x_m}$$

is an open neighbourhood of E. Since the neighbourhoods do not intersect we have that $\mathrm{Cl}\, V \subset U$, as required. Another way to express this result is that if the compact C and the closed E are disjoint subsets of the regular space X then C and E admit disjoint neighbourhoods.

Proposition 6.29

If X is compact regular then X is normal.

For let E, F be disjoint closed sets of X. Then E is compact, since X is compact, and $X - F$ is an open neighbourhood of E. Since X is regular $\mathrm{Cl}\, V \subset S - F$, by Proposition 6.28, for some open neighbourhood V of E. Therefore $X - \mathrm{Cl}\, V$ is an open neighbourhood of F which proves Proposition 6.29.

6.6 Compactly regular spaces

Regularity, as we have seen, means that there exists a closed neighbourhood base at each point. Suppose that we modify this by requiring a compact neighbourhood base at each point. There does not appear to be any settled terminology for this condition unless the Hausdorff property also holds in which case local compactness would be the appropriate term. It is tempting to use the same term for the condition in the non-Hausdorff case but this would be contrary to established practice. The term "compactly regular" seems not inappropriate, and so I propose

Definition 6.30

The topological space X is compactly regular if at each point of X the compact neighbourhoods form a neighbourhood base.

Clearly every discrete space is compactly regular, using the one-point sets as basic neighbourhoods. Closed subspaces of compactly regular spaces are compactly regular but subspaces in general are not. The real line $\mathbb{R}$ is compactly regular, for example, while the rational line $\mathbb{Q}$ is not.

In fact, for a regular space it is only necessary to produce one compact neighbourhood of each point since the closed neighbourhoods form a base, and so the traces of the closed neighbourhoods on the compact neighbourhood form a compact base. In particular, every compact regular space is compactly regular.

Due to the definition of the product topology, the topological product of a family of compactly regular spaces is compactly regular provided all but a finite number of the factors are compact.

It is easy to see that the direct image of a compactly regular space under a continuous open function is also compactly regular. Hence the condition is a topological invariant.

Proposition 6.31

Let $p\colon Y \to Z$ be a quotient map, where Y and Z are topological spaces. Then the product

$$q = \mathrm{id} \times p\colon X \times Y \to X \times Z$$

is a quotient map, for all compactly regular spaces X.

For let $W \subset X \times Z$ be such that $q^{-1}W$ is open and let (x,y) be a point of $q^{-1}W$. Since X is compactly regular there exists a compact neighbourhood C of x in X and an open neighbourhood V_1 of y in Y such that $C \times V_1 \subset q^{-1}W$. Now the second projection $\pi\colon C \times Y \to Y$ is closed, since C is compact; consider the intersection U of $q^{-1}W$ with $C \times Y$. Since U is an open neighbourhood of $C \times p^{-1}(pV_1)$ there exists, by Proposition 4.15, an open neighbourhood V_2 of $p^{-1}(pV_1)$ such that $C \times V_2 \subset U$. Repeating the process we obtain, for $i = 0, 1, \ldots$, an open neighbourhood V_{i+1} of $p^{-1}(pV_i)$ such that $C \times V_{i+1} \subset U$. Then the union V of the open sets V_i is an open neighbourhood of y such that $C \times V \subset U$, hence Int $C \times V$ is an open neighbourhood of (x,y) contained in U and hence in $q^{-1}W$. However, $V = p^{-1}(pV)$ and so pV is open in Z, since p is a quotient map. Therefore Int $C \times pV$ is an open neighbourhood of $(x, p(y))$ contained in W, and so W is open. This proves Proposition 6.31.

The following example, taken from Kelley [10], is illuminating. Let X be a regular Hausdorff space which is not normal. (In fact, the Sorgenfrey plane meets these requirements, as can be shown without great difficulty.) Let E and F be disjoint closed sets of X such that each neighbourhood of E intersects each neighbourhood of F. Consider the complement of the closed set $R = \Delta X \cup (E \times E) \cup (F \times F)$, regarded as a neighbourhood of $E \times F$. The image of this saturated open set under the natural projection

$$X \times X \to (X/R) \times (X/R)$$

is not open. Thus we have an example where the product of two quotient maps is not a quotient map.

Exercises

6.1.   An equivalence relation R on the topological space X is defined so that $\xi R \eta$ if $\mathrm{Cl}\{\xi\} = \mathrm{Cl}\{\eta\}$. Show that the quotient space X/R has the T_0 property: for any two distinct points at least one admits a neighbourhood which does not contain the other.

6.2.   Show that if H is a subset of the T_1 space X then the intersection of all the neighbourhoods of H in X is H itself.

6.3.   In the T_1 space X the intersection of every family of open sets is open. Show that X is discrete.

6.4.   On the closed interval $X = [-1, 1] \subset \mathbb{R}$ let S be the equivalence relation for which the equivalence classes are pairs $\{x, -x\}$ when $-1 < x < 1$, the one-point sets $\{x\}$ when $x = \pm 1$. Show that X/S is T_1 but not Hausdorff.

6.5.   Let A be a closed subspace of the regular space X. Show that A coincides with the intersection of its closed neighbourhoods.

6.6.   Let $\phi \colon X \to Y$ be a function with graph $\Gamma_\phi \subset X \times Y$, where X and Y are topological spaces, with X Hausdorff. Show that ϕ is continuous if either

(i) Γ_ϕ is compact, or
(ii) Γ_ϕ is closed and Y is compact.

6.7.   Let E, F be disjoint closed subsets of the normal space X. Show that there exist disjoint closed neighbourhoods of E, F in X.

6.8.   Let H be a discrete subgroup of the Hausdorff group G. Show that H is closed in G.

6.9.   Let $\{X_j\}$ be a family of regular spaces. Show that the Hausdorff quotient of the topological product $\prod X_j$ is homeomorphic to the topological product of the family of Hausdorff quotients $\{X_j'\}$.

6.10. Show that if every subset of the Hausdorff space X is compact then X is discrete.

6.11. Consider the real line $\mathbb{R}$ with the topology in which a proper subset H is closed if and only if H is closed in the euclidean topology and bounded with respect to the euclidean metric. Show that $\mathbb{R}$ is T_1 but not Hausdorff.

7
Uniform Spaces

7.1 Uniform structures

The second section of this book is concerned with uniform spaces. These are structured sets of a different kind from those we have studied so far. As we shall see in due course, a uniform structure on a given set determines a topological structure on the same set. However, different uniform structures may determine the same topological structure. Moreover, there exist topological structures which cannot be obtained from a uniform structure.

Metrics and topological group structures give rise to uniform structures, as we shall see. A theory which encompasses many of the essentials of both these important classes of spaces is obviously of considerable interest. But what makes uniform spaces important, as much as anything, is that every compact Hausdorff space admits a uniform structure, and that structure is unique.

A uniformity (or uniform structure) on a given set X is a filter on the cartesian square $X \times X$ satisfying certain conditions. The members of the filter are not called relations (although the notation described in Chapter 1 is still used) but *entourages* (or *surroundings*). Specifically, we have

Definition 7.1

A uniformity on a set X is a filter on $X \times X$ consisting of entourages such that:

  (i)   each entourage D contains the diagonal ΔX,

  (ii) if D is an entourage then $E \subset D^{-1}$ for some entourage E,

  (iii) if D is an entourage then $D' \circ D' \subset D$ for some entourage D'.

Note that if D is an entourage then so is D^{-1}; the reason for not stating (ii) in this form will emerge in a moment. Also note that (iii) can be extended, by iteration, to the condition that for each $n = 1, 2, \ldots$ there exists an entourage E such that $E^n \subset D$. In Figure 7.1, the shaded region suggests an entourage for $X = I$.

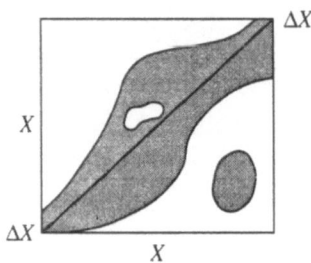

Figure 7.1.

By a *uniform space* we mean a set X together with a uniformity Ω on X; usually X alone is sufficient notation. We also regard the empty set as a uniform space with the empty set as entourage. A *refinement* of the uniformity Ω is a uniformity Ω' on the same set such that each entourage of Ω belongs to Ω'. In this situation we say that Ω' refines Ω or that Ω coarsens Ω'. If the possibility that $\Omega = \Omega'$ is to be excluded we describe the refinement as *strict*. Let us have a few examples.

Definition 7.2

The discrete uniformity on the set X is the uniformity in which every superset of ΔX is an entourage.

In this situation we describe X as a *discrete uniform space*. Clearly the discrete uniformity refines every other uniformity. Going to the other extreme we have

Definition 7.3

The trivial uniformity on the set X is the uniformity in which the full set $X \times X$ is the sole entourage.

In this situation we describe X as a *trivial uniform space*. When X has fewer than two points the discrete uniformity and the trivial uniformity coincide, and no other uniformity is possible. When X has at least two points, however, the discrete uniformity and the trivial uniformity are different, and when X has at least three points there are other possible uniformities as well.

In topology, as we have seen, any family of subsets of a given set X can be used to generate a topology on X, by first taking finite intersections and then taking unrestricted unions. So far as I am aware nothing quite like this is possible in the case of a uniform structure. Given a family of subsets of $X \times X$, each containing the diagonal ΔX, we can always complete the family by taking finite intersections and then pass to the filter thus generated by taking supersets. But at some stage in the process we also have to take the "transitive closure" by taking compositions, in the sense of relation theory, and it is this which causes the difficulty.

The following example may clarify matters. For any set X and any family Γ of subsets of X consider the family $\Gamma^\square$ of subsets of $X \times X$ consisting, for each member M of Γ, of the subset

$$\{(\xi, \eta) : \xi = \eta \text{ or } \xi \in M \text{ or } \eta \in M\}$$

of $X \times X$. Each member of $\Gamma^\square$ is reflexive and symmetric but the composition of any pair of members is just the full set $X \times X$.

The definition (7.1) of the term uniformity includes the word "filter". If we replace this by the term "filter base" then we obtain the definition of what is meant by a base, in this context. Thus a *uniformity base* on a set X is a family of subsets of $X \times X$, also called entourages, such that the same three conditions are satisfied, and which satisfies the conditions for a filter base. The uniformity determined by the base consists of the supersets of the members of the base.

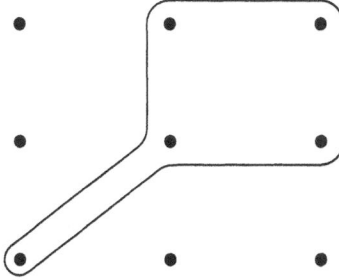

Figure 7.2.

Given a uniformity Ω on X we can, of course, take the whole of Ω as a base, in this sense. More usefully we can take the symmetric entourages of Ω as a base, i.e. the entourages D such that $D = D^{-1}$. In the case of the discrete uniformity the one-member family consisting of the diagonal ΔX constitutes a base. Figure 7.2 shows a base for a uniformity on the point-triple which is neither discrete nor trivial.

In practice, uniformities are usually defined by specifying a base for the uniform structure. For example, consider the real line $\mathbb{R}$. Consider the family of subsets

$$U_\varepsilon = \{(\xi, \eta): |\xi - \eta| < \varepsilon\}$$

of $\mathbb{R} \times \mathbb{R}$, where ε runs through the positive reals. This satisfies the three conditions in Definition 7.1 and constitutes a filter base. The filter generated in this way is called the euclidean uniformity on $\mathbb{R}$. Specifically, a subset D of $\mathbb{R} \times \mathbb{R}$ is an entourage if $U_\varepsilon \subset D$ for some $\varepsilon > 0$. A similar uniformity is defined in the case of $\mathbb{R}^n$.

For another example, consider the set of integers $\mathbb{Z}$. The p-adic structure on $\mathbb{Z}$, for a given prime p, is the uniformity generated by the subsets D_n of $\mathbb{Z} \times \mathbb{Z}$ $(n = 1, 2, \ldots)$, where $(\xi, \eta) \in D_n$ if and only if $\xi \equiv \eta \bmod p^n$. This structure is important in the theory of numbers.

Given a group G there are two ways to associate a relation R_A on G with a subset A of G. One is to write $\xi R_A \eta$ when $\xi^{-1} \cdot \eta \in A$; this is called the left relation. The right relation is defined similarly but with $\eta \cdot \xi^{-1}$ instead of $\xi^{-1} \cdot \eta$. Of course the relations coincide when G is commutative. Note that both left and right relations are equivalence relations in case A is a subgroup of G.

When G is a topological group a uniform structure on G can be defined in two different ways (there are also other ways, but those do not concern us here). Specifically, the left relations determined by the members of the neighbourhood filter at the neutral element e generate the left uniformity, while the right relations generate the right uniformity.

Let us check that the various conditions are satisfied by the family of left relations. Thus let R_U be the (left) relation corresponding to a neighbourhood U of e. Then (i) is satisfied since U contains e, (ii) is satisfied since U contains a neighbourhood V such that $V^{-1} \subset U$, and (iii) is satisfied since U contains a neighbourhood W such that $W \cdot W \subset U$. Moreover, the family $\{R_U\}$ constitutes a filter base since the family $\{U\}$ constitutes a filter, and so all the conditions for a uniformity base are satisfied. The argument for right, rather than left, relations is similar. Note that, in the above construction, if U is confined to a neighbourhood base at the neutral element then the corresponding family of subsets R_U constitutes a base for the left uniformity, and similarly in the case of the right uniformity.

When G is commutative the left and right uniformities coincide, of course. The euclidean uniformity on the real line $\mathbb{R}$ and, more generally, on the real n-space $\mathbb{R}^n$ is an illustration of this, where the neighbourhood base at the neutral element 0 consists of the open ε-balls $U_\varepsilon(0)$. They also coincide when G is discrete, since the entourage given by the neutral element, as a neighbourhood of itself, is just the diagonal ΔG, and so we obtain the discrete uniformity in either case. In general, however, the left and right uniformities need to be distinguished.

For an example where the left and right uniformities are different take G to be the set $\mathbb{R}_* \times \mathbb{R}$ with multiplication

$$(x, y) \cdot (x', y') = (xx', xy' + y).$$

Then G is a topological group with the neighbourhood filter of the neutral element $(1, 0)$ generated by the subsets

$$U_\varepsilon \mathbb{X} = \mathbb{X}\{(x, y) : |x - 1| < \varepsilon, |y| < \varepsilon\},$$

where $\varepsilon > 0$. With

$$\xi = (\varepsilon/4, 0), \qquad \eta = (\varepsilon/4, \varepsilon/8)$$

we have

$$\eta \cdot \xi^{-1} = (1, \varepsilon/2) \in U_\varepsilon, \qquad \xi^{-1} \cdot \eta = (1, 2) \notin U_1.$$

Thus the left relation determined by U_1 does not contain the right relation determined by U_ε for any $\varepsilon > 0$. So the right uniformity is not a refinement of the left uniformity; similarly the left is not a refinement of the right. However, as we shall soon see, this phenomenon cannot occur in the case of compact topological groups.

A metric ρ on a set X determines a uniformity on X, namely the uniformity generated by the family of subsets

$$U_\varepsilon = \rho^{-1}[0, \varepsilon),$$

where ε runs through the positive reals. We refer to U_ε as the ε-entourage determined by ρ. Thus a subset D of $X \times X$ is an entourage, in this uniformity, if there exists a positive ε such that $\rho(\xi, \eta) < \varepsilon$ implies $(\xi, \eta) \in D$. In fact, (i) of Definition 1.22 implies that each entourage contains the diagonal, (ii) implies the symmetry condition, while (iii) implies the condition of weak transitivity.

For example, the discrete metric determines the discrete uniformity. The trivial uniformity on a set with at least two points is not determined by any metric.

Different metrics may determine the same uniformity. For example, the uniformity determined by the metric ρ on the set X is the same as that determined by the metric 2ρ.

7.2 Separated uniformities

If X is a uniform space the intersection R of the entourages constitutes an equivalence relation on X. For R is clearly reflexive and symmetric. To see that R is transitive, let D be an entourage. There exists an entourage E such that $E \circ E \subset D$. Since $R \subset E$ we have $R \circ R \subset E \circ E \subset D$, hence $R \circ R \subset R$. Since $\Delta X \subset R$ we also have $R \subset R \circ R$. Thus $R \circ R = R$, and so R is an equivalence relation, as asserted. To obtain R, of course, it is only necessary to intersect the entourages of a uniformity base. So in the case of a topological group G we see that R is just the subset of $G \times G$ corresponding to the intersection of the neighbourhoods of the neutral element e, in either left or right uniformity, i.e. to the closure of $\{e\}$, since G is regular. In the case of a metric space, R is just the diagonal. This example shows that R itself is not necessarily an entourage.

Returning to the general case we note that the diagonal ΔX is contained in R, since ΔX is contained in every entourage of X; in general, ΔX is a proper subset of R, as the above examples indicate. This gives point to

Definition 7.4

A uniformity on the set X is separated if the diagonal ΔX coincides with the intersection R of the entourages.

For example, the condition is always satisfied in the case of a metric space. In the case of a topological group, with either left or right uniformity, the condition is satisfied if and only if $\{e\}$ constitutes a closed set.

7.3 Totally bounded uniformities[11]

Definition 7.5

The uniform space X is totally bounded if for each entourage D of X there exists a finite subset S of X such that $D[S] = X$.

For example, the trivial uniformity is always totally bounded, while the discrete uniformity is totally bounded if and only if X itself is a finite set.

To understand this important condition better let us introduce a further technical term. Given an entourage D of a uniform space X let us say that a

---

[11] The term precompact is also used instead of the term totally bounded.

subset M of X is *small of order* D, or simply D-*small*, if $M \times M \subset D$. Then the condition in Definition 7.5 is that X can be covered by a finite number of D-small subsets, for each entourage D of X. Clearly it is sufficient for this to be true for the members of a uniformity base. For example let G be a compact topological group. For any neighbourhood U of the neutral element the sets $g \cdot U$, where g runs through the elements of G, form an open covering of G, and this admits a finite subcovering $S \cdot U$, where $S \subset G$ is finite, so that G is totally bounded.

In case the uniformity on X is given by a metric ρ, the condition reduces to the following. By an ε-*net*, where $\varepsilon > 0$, we mean a subset S of X such that the family $\{U_\varepsilon(x) : x \in S\}$ covers X. Here and elsewhere

$$U_\varepsilon(x) = \{\xi \in X : \rho(\xi, x) < \varepsilon\},$$

in keeping with the notation introduced in Chapter 1. So in this case the condition in Definition 7.5 may be expressed as the existence of a finite ε-net for all $\varepsilon > 0$.

If the uniformity on X arises from a metric ρ then X is bounded if X is totally bounded. For take D to be the basic entourage U_1 of X. The condition in Definition 7.5 states that no point of X is further than distance 1 from some point of a finite subset S_1 of X. Since S_1 is finite the diameter diam S_1 is defined, and then $k = 2 + \text{diam } S_1$ is a bound for X.

We have already seen, in Chapter 4, that a metric ρ can be converted into a bounded metric ρ' by formulae such as

$$\rho' = \min(1, \rho), \qquad \rho' = \rho(1 + \rho)^{-1}.$$

The associated uniform structure is not affected by this change. For in the case of the first formula, each basic ε-entourage ($\varepsilon > 0$) in the case of ρ contains the basic ε'-entourage in the case of ρ', where $\varepsilon' = \min(1, \varepsilon)$, while each basic ε'-entourage ($\varepsilon' > 0$) in the case of ρ' contains a basic ε'-entourage in the case of ρ. It follows from these observations that bounded metric spaces are not necessarily totally bounded. For example, the real line $\mathbb{R}$ is not bounded, hence not totally bounded, with the euclidean metric, and so remains not totally bounded after conversion of the metric into a bounded metric.

Proposition 7.6

Let G be a topological group which is totally bounded in either the left or right uniformity. Then the uniformities coincide. In particular they coincide when G is compact.

For suppose that G is totally bounded in the right uniformity (the other case is similar). Given a neighbourhood U of the neutral element there exists a neighbourhood V of the neutral element such that $V \cdot V \cdot V^{-1} \subset U$. Since G is totally bounded we have $V \cdot S = G$ for some finite subset $S = \{g_1, \ldots, g_n\}$ of G. Now $g_j \cdot W_j \cdot g_j^{-1} \subset V$ for some neighbourhood W_j of the neutral element, where $j = 1, \ldots, n$. Then if $W = W_1 \cap \ldots \cap W_n$ and $g \in G$ we have $g \in V \cdot g_j$ for some j and so

$$g \cdot W \cdot g^{-1} \subset W \cdot g_j \cdot W \cdot g_j^{-1} \cdot V^{-1} \subset V \cdot V \cdot V^{-1} \subset U.$$

Thus the union W' of the conjugates $g \cdot W \cdot g^{-1}$ for all $g \in G$ is a neighbourhood of the neutral element such that $g \cdot W' \cdot g^{-1} = W'$. Therefore the left and right uniformities coincide.

7.4 Uniform continuity

In the theory of uniform spaces the structure-preserving functions, in the inverse image sense, are the uniformly continuous functions, as in

Definition 7.7

Let $\phi \colon X \to Y$ be a function, where X and Y are uniform spaces. Then ϕ is uniformly continuous if the inverse image $(\phi \times \phi)^{-1}E$ is an entourage of X for each entourage E of Y.

Thus ϕ is necessarily uniformly continuous if the uniformity of X is discrete or if the uniformity of Y is trivial.

Constant functions are uniformly continuous. For suppose that $\phi X = \{y_0\}$ for some point y_0 of Y. The inverse image of every entourage E of Y with respect to $\phi \times \phi$ is the full set $X \times X$ since $(y_0, y_0) \in \Delta Y \subset E$, and the full set is always an entourage.

Note that the identity function on any uniform space is uniformly continuous. Also that if $\phi \colon X \to Y$ and $\psi \colon Y \to Z$ are uniformly continuous, where X, Y and Z are uniform spaces, then the composition $\psi\phi \colon X \to Z$ is uniformly continuous.

Clearly it is sufficient if the condition in Definition 7.7 is satisfied for all entourages E in a base for the uniformity of Y.

In the metric case, for example, the condition reduces to the following

Condition 7.8

Let $\phi \colon X \to Y$ be a function, where X and Y are metric spaces with metrics ρ, σ, respectively. Then ϕ is uniformly continuous if for each $\varepsilon > 0$ there exists a $\delta > 0$ such that $\rho(\xi, \eta) < \delta$ implies $\sigma(\phi(\xi), \phi(\eta)) < \varepsilon$ for all $\xi, \eta \in X$.

For example, take $X = Y = \mathbb{R}$, with the euclidean uniformity, and take $\phi \colon \mathbb{R} \to \mathbb{R}$ to be the squaring function given by $\phi(x) = x^2$. To see that ϕ is not uniformly continuous, take $\varepsilon = 1$ in the above; then for any $\delta > 0$ we have

$$|\phi(\xi) - \phi(\eta)| \geq 1$$

when $\xi = \delta^{-1} + \frac{1}{2}\delta$, $\eta = \delta^{-1} - \frac{1}{2}\delta$. Almost the same argument shows that real multiplication

$$\mathbb{R} \times \mathbb{R} \to \mathbb{R},$$

is not uniformly continuous, with the euclidean uniformities.

A point to note is that for any metric space X the metric $\rho \colon X \times X \to \mathbb{R}$ itself is necessarily uniformly continuous, using the euclidean metric on $\mathbb{R}$. In fact, if $D = U_{\varepsilon/2}$ $(\varepsilon > 0)$ is a basic entourage of X then for $(\xi_i, \eta_i) \in D$ $(i = 1, 2)$ we have

$$|\rho(\xi_1, \eta_1) - \rho(\xi_2, \eta_2)| < \varepsilon,$$

by the triangle inequality; therefore ρ is uniformly continuous, as asserted.

Another straightforward consequence of the definition is

Proposition 7.9

Let $\phi \colon G \to H$ be a (continuous) homomorphism, where G and H are topological groups. Then ϕ is uniformly continuous, with respect to both left and right uniformities.

For if E is a basic entourage of H, corresponding to a neighbourhood N of the neutral element in H, then $(\phi \times \phi)^{-1}E$ is a basic entourage of G, corresponding to the neighbourhood $\phi^{-1}N$ of the neutral element in G.

Corollary 7.10

Let G be a commutative topological group. Then the inversion function $G \to G$ and the multiplication function $G \times G \to G$ are uniformly continuous.

The same conclusions hold, of course, when G is discrete. For general G, however, all one can say is that inversion is uniformly continuous, in either left or right uniformity, if and only if the left and right uniformities coincide.

7.5 Uniform equivalences

Definition 7.11

A uniform equivalence $\phi\colon X \to Y$, where X and Y are uniform spaces, is a bijective function such that both ϕ and ϕ^{-1} are uniformly continuous.

In particular, if X and Y are metric spaces and ϕ is a distance-preserving bijection then ϕ is a uniform equivalence. In such a case we also describe ϕ as an *isometry*. The orthogonal transformations of $\mathbb{R}^n$ are examples of isometries.

It is clear that the existence of a uniform equivalence constitutes an equivalence relation between uniform spaces. A property of uniform spaces which is invariant under uniform equivalence is called a *uniform invariant*. Total boundedness is an example of a uniform invariant, in view of

Proposition 7.12

Let $\phi\colon X \to Y$ be a uniformly continuous surjection, where X and Y are uniform spaces. If X is totally bounded then so is Y.

For let E be an entourage of Y. Then $D = (\phi \times \phi)^{-1}E$ is an entourage of X, since ϕ is uniformly continuous. Since X is totally bounded there exists a finite subset S of X such that $D[S] = X$. Then $Y = E[\phi S]$, where ϕS is finite, and so Y is totally bounded, as asserted.

Note that if $\phi\colon X \to Y$ is a bijection, where X is a set and Y is a uniform space, we can always impose a uniformity on X so as to make ϕ a uniform equivalence. Specifically, the entourages of X are precisely the inverse images, under $\phi \times \phi$, of the entourages of Y.

Definition 7.13

The uniform space X is uniformly homogeneous if for each pair of points x, x' of X there exists a uniform equivalence $\phi\colon X \to X$ such that $\phi(x) = x'$.

For example, discrete uniform spaces and trivial uniform spaces are uniformly homogeneous.

For another example, consider a topological group G, with the left uniformity. For each element g of G the left translation $g_\#\colon G \to G$ is uniformly continuous, since $(g_\# \times g_\#)D = D$ for each basic entourage D of G. Since the inverse of $g_\#$ is just $(g^{-1})_\#$ it follows at once that $g_\#$ is a uniform equivalence, and hence that G is uniformly homogeneous. Similarly, for right translation, with the right uniformity.

7.6 The uniform product

Just as the cartesian product of topological spaces can be given a topology so the cartesian product of uniform spaces can be given a uniformity. Specifically, let $\{X_j\}$ be a family of uniform spaces. Disregarding the uniformities for a moment consider the cartesian product $X = \prod X_j$. We identify $X \times X$ with the cartesian product $\prod(X_j \times X_j)$ by the obvious rearrangement of factors. This rearrangement being understood we regard the cartesian product $\prod D_j$, where $D_j \subset X_j \times X_j$ for each index j, as a subset of $X \times X$. In particular, we regard the restricted product set $\prod D_j$, where D_j is full for all but a finite number of indices j, as a subset of $X \times X$. Now consider the family of subsets of $X \times X$ consisting of the restricted product sets $\prod D_j$ such that D_j is an entourage of X_j, for each index j. This family constitutes a base for a uniformity on X, called the *product uniformity*, and X, with this uniformity, is called the *uniform product* of the members of the family $\{X_j\}$. Note that if the uniformity on X_j is generated by a base, for each index j, then the product uniformity on $\prod X_j$ is generated by the family of restricted product sets $\prod D_j$ where D_j is basic or full for each j.

It follows at once from the definition of the product uniformity that each of the projections π_j is uniformly continuous. In fact, the product uniformity may be described as the coarsest uniformity on the cartesian product for which this is true. As a result the uniform product is characterized by the following property:

Proposition 7.14

Let $\{X_j\}$ be a family of uniform spaces. Let $\phi: A \to \prod X_j$ be a function, where A is a uniform space and $\prod X_j$ is the uniform product. Then ϕ is uniformly continuous if and only if each of the functions $\phi_j = \pi_j\phi: A \to X_j$ is uniformly continuous.

As far as possible, one should try and deduce results about the uniform product from Proposition 7.14, rather than go back to the definition of the product uniformity. For example, let $\{\psi_j\}$ be a family of uniformly continuous functions $\psi_j: X \to Y_j$, where X_j and Y_j are uniform spaces. Then it follows at once from Proposition 7.14 that the product function

$$\prod \psi_j: \prod X_j \to \prod Y_j$$

is uniformly continuous, since the jth component of $\prod \psi_j$ is the uniformly continuous function $\psi_j\pi_j$. Notice that if each of the ψ_j is a uniform equivalence then so is the product $\prod \psi_j$. It follows that the uniform product of uniformly homogeneous uniform spaces is again uniformly homogeneous.

For another illustration, consider the uniform product X^J of J copies of the uniform space X, for some indexing set J. Then Proposition 7.14 shows that the diagonal function

$$\Delta \colon X \to X^J$$

is uniformly continuous, since each component of Δ is the identity function on X.

The proofs of the following two results are straightforward and will be left to serve as exercises.

Proposition 7.15

Let $\{X_j\}$ be a family of separated uniform spaces. Then the uniform product $\prod X_j$ is separated.

Proposition 7.16

Let $\{X_j\}$ be a family of totally bounded uniform spaces. Then the uniform product $\prod X_j$ is totally bounded.

Let $\{G_j\}$ be a family of topological groups. Each member G_j of the family can be regarded as a uniform space, using the left uniformity, and then the cartesian product $\prod G_j$ can be regarded as a uniform space, using the product uniformity. However, $\prod G_j$ can also be regarded as a topological group, using the direct product group structure, and hence as a uniform space, using the left uniformity. It is easy to check that these uniformities on $\prod G_j$ coincide. Similarly, using right uniformity instead of left uniformity, throughout.

Let us pause in the development of the theory for a little while and consider some more specific examples. To start with it follows from Corollary 7.10 that the addition function $\mathbb{R} \times \mathbb{R} \to \mathbb{R}$ is uniformly continuous, in the euclidean uniformity, likewise the negative $\mathbb{R} \to \mathbb{R}$. It is also easy to see that dilatations as well as translations are uniformly continuous, and hence that the affine transformation $p \colon \mathbb{R} \to \mathbb{R}$ given by

$$p(\xi) = \lambda \xi + \mu \qquad (\xi \in \mathbb{R})$$

is uniformly continuous for all real λ, μ. Other uniformly continuous functions $\mathbb{R} \times \mathbb{R} \to \mathbb{R}$ include max and min; the proofs are straightforward.

A similar situation obtains in higher dimensions. Thus the negative function $\mathbb{R}^n \to \mathbb{R}^n$ and the addition function $\mathbb{R}^n \times \mathbb{R}^n \to \mathbb{R}^n$ are uniformly continuous, in the euclidean uniformity on $\mathbb{R}^n$ ($n \geq 1$). Moreover, the affine transformation $p \colon \mathbb{R}^n \to \mathbb{R}^n$ given by

$$p(\xi) = \lambda \xi + \mu \qquad (\xi \in \mathbb{R}^n)$$

is uniformly continuous, where λ is a real $n \times n$ matrix and $\mu \in \mathbb{R}^n$. From these basic facts we can draw conclusions about real-valued functions with a uniform space X as domain. Thus if $\phi \colon X \to \mathbb{R}$ is uniformly continuous then so, by postcomposition with the negative function $\mathbb{R} \to \mathbb{R}$, is $-\phi \colon X \to \mathbb{R}$. Again, if $\phi, \psi \colon X \to \mathbb{R}$ are uniformly continuous then so are

$$\phi + \psi, \ \max(\phi, \psi), \ \min(\phi, \psi) \colon X \to \mathbb{R}.$$

To see this, all we need to do is observe that $\phi + \psi$ can be expressed as the composition

$$X \xrightarrow{\Delta} X \times X \xrightarrow{\phi \times \psi} \mathbb{R} \times \mathbb{R} \to \mathbb{R},$$

where the last function is given by addition, and similarly with max and min.

In the case of the punctured line $\mathbb{R}_* = \mathbb{R} - \{0\}$ the obvious uniformity to use is the one determined by the multiplicative group structure. This has the property that the operations

$$\mathbb{R}_* \to \mathbb{R}_*, \qquad \mathbb{R}_* \times \mathbb{R}_* \to \mathbb{R}_*$$

of inversion and multiplication are uniformly continuous.

With the multiplicative uniformity, the inverse (reciprocal) ϕ^{-1} of a uniformly continuous function $\phi \colon X \to \mathbb{R}_*$ is uniformly continuous, likewise the product $\phi \cdot \psi$ of uniformly continuous functions $\phi, \psi \colon X \to \mathbb{R}_*$ is uniformly continuous, for each uniform space X.

7.7 The induced uniformity

We now return to the development of the theory with

Definition 7.17

Let $\phi \colon X \to Y$ be a function, where X and Y are uniform spaces. The uniformity on X is induced by ϕ from the uniformity on Y if the uniformity on X is generated by the inverse images, with respect to $\phi \times \phi$, of the entourages of Y.

The condition here is stronger than uniform continuity since X is not allowed to have entourages other than those which arise from Y in the prescribed manner.

The situation may occur in which we are given a function $\phi \colon X \to Y$, where Y is a uniform space and X is a set. Then we can use the procedure indicated in Definition 7.17 to give X a uniformity. This uniformity, which is called the induced uniformity, may be described as the coarsest uniformity such that ϕ is uniformly continuous. For example, if Y has the trivial uniformity then so does X.

The induced uniformity is transitive in the following sense. Let $\phi\colon X \to Y$ and $\psi\colon Y \to Z$ be functions, where X, Y and Z are uniform spaces. If Y has the uniformity induced by ψ from the uniformity of Z and X has the uniformity induced by ϕ from the uniformity of Y then X has the uniformity induced by $\psi\phi$ from the uniformity of Z. The proof is obvious.

The following property is characteristic of the induced uniformity.

Proposition 7.18

Let $\phi\colon X \to Y$ and $\psi\colon Y \to Z$ be functions, where X, Y and Z are uniform spaces. Suppose that Y has the uniformity induced by ψ from the uniformity of Z. Then ϕ is uniformly continuous if (and only if) $\psi\phi$ is uniformly continuous.

For if E is an entourage of Y then $E \supset (\psi \times \psi)^{-1}F$, for some entourage F of Z, since Y has the induced uniformity. Then

$$(\phi \times \phi)^{-1}E \supset (\phi \times \phi)^{-1}(\psi \times \psi)^{-1}F = (\psi\phi \times \psi\phi)^{-1}F,$$

which is an entourage of X when $\psi\phi$ is uniformly continuous.

The special case when the function is injective is particularly important.

Definition 7.19

Let $\phi\colon X \to Y$ be an injection, where X and Y are uniform spaces. Then ϕ is a uniform embedding if the entourages of X are precisely the inverse images under $\phi \times \phi$ of the entourages of Y.

A uniformly continuous injection is necessarily a uniform embedding when the domain has trivial uniform structure, for example, when the domain is a one-point space. However, in general, the domain of a uniformly continuous injection will have entourages which are not inverse images of entourages of the codomain. For example, take the identity function on any set with the uniformity of the domain a strict refinement of the uniformity of the codomain.

Of course, Proposition 7.18 and other results about induced uniformities apply to uniform embeddings as a special case. In addition, we have

Proposition 7.20

Let $\phi\colon X \to Y$ and $\psi\colon Y \to Z$ be uniformly continuous functions, where X, Y and Z are uniform spaces. If $\psi\phi$ is a uniform embedding then so is ϕ.

For then each entourage of X is of the form $(\psi\phi \times \psi\phi)^{-1}F$, where F is an entourage of Z, and so is of the form $(\phi \times \phi)^{-1}(\psi \times \psi)^{-1}F$, where $(\psi \times \psi)^{-1}F$ is an entourage of Y.

In particular, take $Z = X$ and $\psi\phi$ the identity. In that case ψ is called a *left inverse* of ϕ and Proposition 7.20 states that a uniformly continuous function is a uniform embedding if it admits a (uniformly continuous) left inverse. Thus we obtain

Corollary 7.21

Let $\phi: X \to Y$ be a uniformly continuous function, where X and Y are uniform spaces. Then the graph function

$$\Gamma_\phi: X \to X \times Y$$

is a uniform embedding.

Here Γ_ϕ may be expressed as the composition

$$X \xrightarrow{\Delta} X \times X \xrightarrow{\text{id} \times \phi} X \times Y;$$

the left inverse is given by projecting onto the first factor. Similarly, we obtain

Corollary 7.22

For any uniform space X and indexing set J the diagonal function

$$\Delta: X \to X^J$$

is a uniform embedding.

Clearly the condition in Definition 7.19 will be satisfied, in general, if it is satisfied by the members of a base for the uniformity of the domain. Hence we obtain

Proposition 7.23

Let $\{\phi_j\}$ be a family of uniform embeddings $\phi_j: X_j \to Y_j$, where X_j and Y_j are uniform spaces. Then the uniform product

$$\prod \phi_j: \prod X_j \to \prod Y_j$$

is a uniform embedding.

Suppose now that we have a uniform space X and a subset A of X. For each subset D of $X \times X$ the trace $(A \times A) \cap D$ on $A \times A$ is just the inverse image of D under the injection $A \times A \to X \times X$. There is a unique way to impose a uniformity on A so as to make the injection $A \to X$ a uniform embedding. Specifically, we take the entourages of A to be the traces on $A \times A$ of the entourages of X, as illustrated in Figure 7.3. This is called the *relative uniformity* and A, with this uniformity, is called a (uniform) subspace of X.

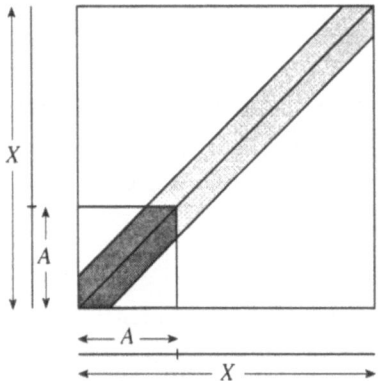

Figure 7.3. Trace of an entourage on a subspace.

It is important to appreciate that if A is a proper subset of X then entourages of A, in the relative uniformity, cannot be entourages of X, since they do not contain ΔX.

It should be noted that the multiplicative uniformity on the punctured real line $\mathbb{R}_*$ is not the same as the euclidean uniformity which $\mathbb{R}_*$ obtains as a subset of $\mathbb{R}$. Nevertheless, the exponential function constitutes a uniform equivalence between $\mathbb{R}$, with the euclidean uniformity, and $\mathbb{R}_+^* = (0, \infty)$, with the multiplicative uniformity.

Some of the results we have proved for uniform embeddings are particularly useful in relation to subspaces. Thus Proposition 7.18 shows that if $\phi \colon X \to Y$ is uniformly continuous, where X and Y are uniform spaces, then so is the function $\phi' \colon X' \to Y'$ determined by ϕ for each subspace X' of X and each subspace Y' of Y which contains $\phi X'$. Moreover, Proposition 7.20 shows that if, further, ϕ is a uniform embedding then so is ϕ'. Consequently, a uniform embedding $\phi \colon X \to Y$ maps X by a uniform equivalence onto the subspace ϕX of Y; more generally, ϕ determines a uniform equivalence $X' \to \phi X'$ for every subspace X' of X. Taking $X = Y = \mathbb{R}$, in particular, with ϕ an affine transformation we deduce that all the open intervals (α, β), where $\alpha < \beta$, are uniformly equivalent, likewise all the closed intervals $[\alpha, \beta]$ are uniformly equivalent, with respect to the euclidean uniformity.

Notice that a base for the uniformity of a uniform space X determines, by taking traces, a base for the relative uniformity of each subspace of X. From this it follows that the relative uniformity is compatible with the uniform product. Specifically, let $\{X_j\}$ be a family of uniform spaces and let A_j be a subspace of X_j for each index j. Then the product uniformity on $\prod A_j$ coincides with the relative uniformity obtained from the uniform product $\prod X_j$ by Proposition 7.23.

Proposition 7.24

Let G' be a subgroup of the topological group G. The uniformity on G' determined by its topological group structure coincides with the relative uniformity obtained from G, as a uniform space.

The proof is straightforward and will be left as an exercise. Instead we prove two rather less straightforward results.

Proposition 7.25

Let X be a totally bounded uniform space. Then each subspace A of X is also totally bounded.

For let E be an entourage of A. Then E is the trace on $A \times A$ of an entourage D of X. Let D' be a symmetric entourage such that $D' \circ D' \subset D$. Since X is totally bounded there is a finite subset $S = \{x_1, \ldots, x_n\}$ of X such that $D'[S] = X$. Let $S' \subset S$ be the subset of those x_i ($i = 1, \ldots, n$) such that $D'[x_i]$ intersects A and for each x_i in S', choose a point $a_i \in D'[x_i] \cap A$. Since D' is symmetric $x_i \in D'[a_i]$, and so

$$D'[x_i] \subset D' \circ D'[a_i] \subset D[a_i].$$

Let T be the set of a_i's. Then $A \subset D'[S'] \subset D[T]$, and so $A = E[T]$, as required.

Proposition 7.26

Let $\{X_j\}$ be a finite covering of the uniform space X. Suppose that X_j is totally bounded for each index j. Then X is totally bounded.

For let D be an entourage of X. Then $D_j = D \cap (X_j \times X_j)$ is an entourage of X_j. Since X_j is totally bounded there exists a finite subset S_j of X_j such that $D_j[S_j]$ is the full set X_j. Since the covering is finite the union S of the S_j is still a finite subset of X. If $\xi \in X$ then $\xi \in X_j$ for some j and so $(x, \xi) \in D_j$ for some $x \in S_j$. Then $x \in S$ and $(x, \xi) \in D$, as required.

7.8 The coinduced uniformity?

At this point the reader might well expect to find a discussion of quotient uniformities and, more generally, coinduced uniformities. Unfortunately, there are unavoidable difficulties in carrying out anything like the same programme as in the topological case. The trouble lies with weak transitivity, the third uniformity axiom, which tends to be destroyed when identifications are made.

To avoid this trouble some condition is needed which is straightforward to check in practice, such as the following.

Condition 7.27

The equivalence relation R on the uniform space X is compatible with the uniformity if for each entourage D of X there exists an entourage D' of X such that $R \circ D' \subset D \circ R$.

Later I will try and explain where this compatibility condition comes from, but first let us see what properties it has and then discuss some examples. The condition implies that the images of X under $\pi \times \pi$ form a base for a uniformity on the quotient set X/R, such that the natural projection π is uniformly continuous. I refer to this as the *quotient uniformity* and to X/R, with this structure, as the *quotient uniform space*. Note that the condition can be replaced by $D' \circ R \subset R \circ D$, by reversing both sides of the relation. Also that it is sufficient for it to be satisfied for the basic entourages when the uniformity of X is given by a base.

Before continuing with the theory let us examine a special case which is of considerable importance. As we have seen earlier, for any uniformity on a set X the intersection R of the entourages constitutes an equivalence relation on X. In this case if D is an entourage of X we have $\Delta X \subset R \subset D$ and so $R \circ D \subset D \circ R$. Thus the compatibility condition is satisfied. Moreover the intersection of the sets $(\pi \times \pi)D$, as D runs through the entourages of X, is precisely the diagonal of X/R, which is therefore separated. We refer to X/R as the *separated quotient space* associated with X.

Example 7.28

Let G be a topological group and let $\mathrm{Cl}\,\{e\}$ be the closure of the neutral element. Then the factor group $G/\mathrm{Cl}\,\{e\}$, as a uniform space, is precisely the separated quotient space associated with G, as a uniform space.

In the theory of topological spaces, as we have seen, there is the useful concept of open function. In the theory of uniform spaces the corresponding concept is that of uniformly open function.

Definition 7.29

The function $\phi\colon X \to Y$, where X and Y are uniform spaces, is uniformly open if for each entourage D of X there exists an entourage E of Y such that

$$E[\phi(x)] \subset \phi(D[x])$$

for all $x \in X$.

Of course it is sufficient for the condition to be satisfied for basic D when the uniformity of X is given by a base. Thus the natural projections from the uniform product onto its factors are uniformly open. Note that every function ϕ is uniformly open when Y has the discrete uniformity. In the metric case the condition reduces to the following

Condition 7.30

Let $\phi\colon X \to Y$ be a function, where X and Y are metric spaces with metrics ρ, σ respectively. Then ϕ is uniformly open if for each $\delta > 0$ there exists an $\varepsilon > 0$ such that $\sigma(\phi(x), \eta) < \varepsilon$, where $x \in X$ and $\eta \in Y$, implies that $\rho(x, \xi) < \delta$ for some $\xi \in X$ such that $\phi(\xi) = \eta$.

Note that a bijection is uniformly open if and only if its inverse is uniformly continuous. Thus a uniformly continuous bijection is uniformly open if and only if it is a uniform equivalence.

The characteristic property of uniformly open functions is as follows.

Proposition 7.31

Let $\phi\colon X \to Y$ and $\psi\colon Y \to Z$ be functions, where X, Y and Z are uniform spaces. Suppose that ϕ is a uniformly continuous and uniformly open surjection. Then ψ is uniformly continuous if and only if $\psi\phi$ is uniformly continuous.

In one direction this is obvious, since the composition of uniformly continuous functions is uniformly continuous. To prove Proposition 7.31 in the other direction, suppose that $\psi\phi$ is uniformly continuous. If F is an entourage of Z then the inverse image $D = (\psi\phi \times \psi\phi)^{-1}F$ is an entourage of X. Since ϕ is uniformly open there exists an entourage E of Y such that $E[\phi(x)] \subset \phi D[x]$ for all $x \in X$. So if $\phi(\xi) \in E[\phi(x)]$ for some $\xi \in X$ then $\phi(\xi) = \phi(\eta)$ for some $\eta \in D[x]$ and then

$(\psi\phi(x), \psi\phi(\eta)) \in F$. Since $\psi\phi(\xi) = \psi\phi(\eta)$ this shows that ψ is uniformly continuous, as required.

Our next result helps to explain and motivate the compatibilty condition.

Proposition 7.32

Let R be an equivalence relation on the uniform space X. If R is compatible with the uniformity on X and X/R has the quotient uniformity then the projection $\pi\colon X \to X/R$ is both uniformly continuous and uniformly open. Conversely if X/R admits a uniformity such that π is uniformly continuous and uniformly open then R is compatible with the uniformity on X and the uniformity on X/R is the quotient uniformity.

For suppose that R is compatible with the uniformity on X. Given a symmetric entourage D of X there exists a symmetric entourage D' of X such that $D' \circ R \subset R \circ D$. Then

$$(\pi \times \pi)D'[\pi(x)] = \pi((D' \circ R)[x]) \subset \pi((R \circ D)[x]) = \pi(D[x]),$$

for all $x \in X$. Since $(\pi \times \pi)D'$ is an entourage of X/R in the quotient uniformity this shows that π is uniformly open.

Conversely suppose that π is uniformly continuous and uniformly open, with respect to some uniformity on X/R. If D is a symmetric entourage of X then, since π is uniformly open, we have

$$E[\pi(x)] \subset \pi D[x] \qquad (x \in X)$$

for some symmetric entourage E of X/R, and so $(\pi \times \pi)D$ is an entourage of X/R since $E \subset (\pi \times \pi)D$. On the other hand if E is an entourage of X/R then $D = (\pi \times \pi)^{-1}E$ is an entourage of X, by uniform continuity, and then $E = (\pi \times \pi)D$. This completes the proof.

Corollary 7.33

Let R be a compatible equivalence relation on the uniform space X, and let X/R be the quotient uniform space. Then for any uniform space Y a function $\phi\colon X/R \to Y$ is uniformly continuous if and only if $\phi\pi\colon X \to Y$ is uniformly continuous.

Proposition 7.34

Let R be a compatible equivalence relation on the uniform space X. Suppose that X/R is totally bounded. Also suppose that all the equivalence classes $R[x]$ are totally bounded. Then X is totally bounded.

For let D be an entourage of X. Then $F \circ R \subset R \circ E \subset D \circ R$ for some symmetric entourages E, F of X. Since X/R is totally bounded there exists a finite subset S of X such that

$$X/R = (\pi \times \pi)F[\pi S] = \pi((F \circ R)[S]) \subset \pi(E[S]),$$

whence $X \subset (R \circ E)[S] \subset (D \circ R)[S]$. Since $R[S]$ is a totally bounded subset of X, by Proposition 7.25, there exists a finite subset T of X such that $R[S] \subset D[T]$, where $X \subset (D \circ D)[T]$.

Let H be a subgroup of the topological group G, with the right uniformity. Since the equivalence relation on G for which left cosets are the equivalence classes is compatible with the uniformity, the natural projection $\pi \colon G \to G/H$ is uniformly open, with respect to the right quotient uniformity. Hence the (left) translations of G/H by elements of G are uniform equivalences.

The various uniformly invariant general properties of uniform spaces can be translated into the appropriate form for these quotient uniform spaces. Thus the condition for G/H to be separated is that the intersection of the subsets $W \cdot H$, for all neighbourhoods W of the neutral element, reduces to H. Also the condition for G/H to be totally bounded is that for each such W there exists a finite subset S of G such that $W \cdot S \cdot H = G$.

Since the natural projection from G to G/H is uniformly open it follows that the natural projection from G/K to G/H is uniformly open, where K is any subgroup of H. Hence we obtain

Proposition 7.35

Let G be a topological group and let $K \subset H \subset G$ be subgroups. Suppose that G/H and H/K are totally bounded in the right quotient uniformity. Then G/K is totally bounded in the right quotient uniformity.

In case H is a normal subgroup of G the factor group G/H is defined, as a topological group (there is no need to distinguish between left and right cosets in this case). The natural projection $\pi \colon G \to G/H$ is a continuous homomorphism, therefore uniformly continuous.

Note that if $\phi \colon G \to G'$ is an open continuous epimorphism, where G and G' are topological groups, then ϕ is uniformly open. Hence the induced function $\phi \pi^{-1} \colon G/H \to G'$ is a topological isomorphism, where $H = \ker(\phi)$.

Exercises

7.1. Does the family
$$\{(\xi, \eta): |\xi - \eta| \in \mathbb{Q}, |\xi - \eta| < \varepsilon\}$$
of subsets of $\mathbb{R} \times \mathbb{R}$, where ε runs through positive reals, generate a uniform structure on $\mathbb{R}$?

7.2. Show that a relation R on the set X forms a uniformity base on X if and only if R is an equivalence relation.

7.3. Let ϕ be a real-valued function which is uniformly continuous on the uniform space X. Suppose that $|\phi(\xi)| \geq \varepsilon$ for some $\varepsilon > 0$ and all $\xi \in X$. Show that $1/\phi$ is uniformly continuous.

7.4. Show that the exponential function $\mathbb{R} \to \mathbb{R}$ is not uniformly continuous, with the euclidean uniformity.

7.5. Show that the uniform product of infinitely many discrete uniform spaces is not, in general, discrete.

7.6. The subspace A of the uniform space X is said to be bounded if for each entourage D of X there exists an integer n such that $A \subset D^n[S]$ for some finite subset S of A. Show that the union of a pair of bounded subspaces is bounded.

7.7. Let R be a compatible equivalence relation on the uniform space X, and let A be a subset of X which is saturated with respect to R, in the sense that $R[A] \subset A$. Show that the equivalence relation $S = R \cap (A \times A)$ on A is compatible with the relative uniform structure.

7.8. The real-valued function ρ is defined so that $\rho(\xi, \eta)$ is the shortest geodesic distance between the points ξ and η of the sphere S^2. Show that ρ is a metric on S^2 and that the uniform structure determined by ρ is the same as that determined by the euclidean metric.

7.9. On the real line the metric ρ_n, for n odd, is given by
$$\rho_n(\xi, \eta) = |\xi^n - \eta^n| \qquad (\xi, \eta \in \mathbb{R}).$$
Show that the uniform structures determined by these metrics are all distinct.

7.10. Let ϕ, ψ be real-valued functions which are uniformly continuous on the uniform space X. Show that if ϕ and ψ are bounded then their product $\phi \cdot \psi$ is uniformly continuous.

7.11. Let $\phi: X \to Y$ and $\psi: Y \to Z$ be functions, where X, Y and Z are uniform spaces.

(i) Suppose that $\psi\phi$ is uniformly open and ϕ is a uniformly continuous surjection. Show that ψ is uniformly open.

(ii) Suppose that $\psi\phi$ is uniformly open and that ψ is a uniformly continuous injection. Show that ϕ is uniformly open.

8

The Uniform Topology

8.1 Uniform neighbourhoods

We begin this chapter by showing that each uniformity on a given set determines a topology, on the same set.

Definition 8.1

Let X be a uniform space. The uniform topology on X is the topology in which a neighbourhood base at a point ξ of X is formed by the family of sets $D[\xi]$, where D runs through the entourages of X.

We have to check that the system of neighbourhood bases thus defined is coherent, in the sense of Definition 1.20. Given the basic neighbourhood $D[\xi]$ of ξ, choose an entourage D' such that $D' \circ D' \subset D$. Then $D'[\eta] \subset D[\xi]$ whenever $\eta \in D'[\xi]$, as required. Thus the system of neighbourhood bases is coherent and so determines a topology. We therefore refer to $D[\xi]$ as a *uniform neighbourhood* of ξ. Also we refer to $D[H]$ as a *uniform neighbourhood* of H for each subset H of X.

Note that if instead of using all the entourages we simply let D run through the members of a base for the uniformity the system of neighbourhood bases thus defined remains coherent and determines the same topology.

For example, suppose that X has the discrete uniformity, for which the diagonal $\Delta X = \Delta$ constitutes a base; since $\Delta[x] = \{x\}$ for all x, the associated topology is discrete. For another example, suppose that X has the trivial

127

uniformity, for which the full set $X \times X$ is the sole entourage and so X is the sole uniform neighbourhood; the associated topology is therefore trivial.

In the case of a metric space X the neighbourhood base at each point ξ determined by the uniformity base of ε-entourages ($\varepsilon > 0$) is precisely that determined by the family of open ε-balls, since

$$U_\varepsilon[\xi] = U_\varepsilon(\xi).$$

So in this case the uniform topology is precisely the metric topology. In particular, the uniform topology determined by the euclidean uniformity on $\mathbb{R}^n$ is precisely the euclidean topology.

Not every topology on a given set can be obtained as the uniform topology of a uniformity; a necessary and sufficient condition will be obtained in Chapter 11. Moreover, if a topology can be so obtained there may be more than one uniformity which gives rise to the topology. For example, let X be an infinite set. With each partition $\{H_j\}$ of X we can associate the superset

$$\bigcup_j (H_j \times H_j)$$

of the diagonal in $X \times X$, as indicated in Figure 8.1. For unrestricted partitions these supersets generate the discrete uniformity. For finite partitions they generate a different uniformity, but the associated uniform topology is still the discrete topology.

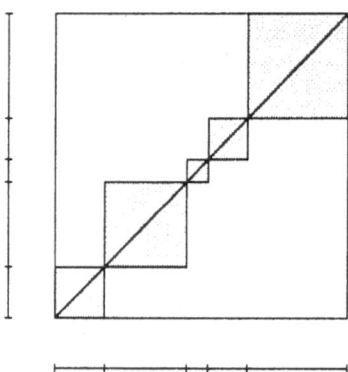

Figure 8.1. Basic entourage determined by partition.

Proposition 8.2

Let $\phi: X \to Y$ be uniformly continuous, where X and Y are uniform spaces. Then ϕ is continuous, with respect to the uniform topologies.

For let x be a point of X. Consider the uniform neighbourhood $E[\phi(x)]$ of $\phi(x)$ in Y, where E is an entourage of Y. The inverse image $(\phi \times \phi)^{-1}E = D$ is an entourage of X. Then $D[x]$ is a neighbourhood of x such that $\phi D[x] \subset E[\phi(x)]$ and so ϕ is continuous at x, as required. Similarly, we have

Proposition 8.3

Let $\phi\colon X \to Y$ be uniformly open, where X and Y are uniform spaces. Then ϕ is open, with respect to the uniform topologies.

Corollary 8.4

Let $\phi\colon X \to Y$ be a uniform equivalence, where X and Y are uniform spaces. Then ϕ is a homeomorphism with respect to the uniform topologies.

It follows, in particular, that if the uniform space X is homogeneous in the uniform sense then X, with the uniform topology, is homogeneous in the topological sense.

Proposition 8.5

Let G be a topological group. Then the uniform topology determined by the left or right uniformity on G coincides with the original topology.

For if N is a neighbourhood of the neutral element e in the original topology then $N = D[e]$, where D is the entourage corresponding to N in either uniformity. Thus e has the same neighbourhood filter in the uniform topology as in the original topology. The general case follows by using translation.

Proposition 8.6

Let $\phi\colon X \to Y$ be a function, where X and Y are uniform spaces. If the uniformity of X is induced by ϕ from the uniformity of Y then the uniform topology of X is induced by ϕ from the uniform topology of Y.

For consider the uniform neighbourhood $D[x]$ of x in X, where D is an entourage of X. If X has the induced uniformity then $D = (\phi \times \phi)^{-1}E$ for some entourage E of Y. Then $D[x] = \phi^{-1}(E[\phi(x)])$, where $E[\phi(x)]$ is a uniform neighbourhood of $\phi(x)$ in Y, and so X has the induced topology, as asserted. In particular, if ϕ is an embedding in the uniform sense then ϕ is an embedding in the topological sense and we obtain

Corollary 8.7

Let A be a subspace of the uniform space X. Then the uniform topology on A obtained from the relative uniformity on A coincides with the relative topology on A obtained from the uniform topology on X.

Similarly, we have

Proposition 8.8

Let $\{X_j\}$ be a family of uniform spaces. Then the uniform topology on the cartesian product $\prod X_j$ obtained from the product uniformity coincides with the product topology obtained from the associated family $\{X_j\}$ of topological spaces.

Proposition 8.9

Let R be a compatible equivalence relation on the uniform space X. Then the uniform topology obtained from the quotient uniformity on X/R coincides with the quotient topology obtained from the uniform topology on X.

Recall from the previous chapter that a uniformity is said to be separated if the diagonal is the intersection of the family of entourages. We prove

Proposition 8.10

Let X be a uniform space. The uniformity is separated if and only if the uniform topology has the Hausdorff property.

For let ξ, η be distinct points of X. If the uniformity is separated then $(\xi, \eta) \notin D$ for some entourage D. Choose a symmetric entourage D' such that $D' \circ D' \subset D$. Then $D'[\xi]$ and $D'[\xi]$ provide neighbourhoods of ξ and η which are disjoint.

Conversely, suppose that X is Hausdorff, in the uniform topology. Then there exist uniform neighbourhoods $D_\xi[\xi]$ of ξ and $D_\eta[\eta]$ of η which are disjoint. There $D_\xi \cap D_\eta$ is an entourage, since D_ξ and D_η are entourages, and $(\xi, \eta) \notin D_\xi \cap D_\eta$. Thus the uniformity is separated, which completes the proof.

8.2 Closure in uniform spaces

Proposition 8.11

Let X be a uniform space. For each symmetric entourage D of X and each subset M of $X \times X$ the subset $D \circ M \circ D$ is a neighbourhood of M in the topological product $X \times X$. Moreover, the closure of M is given by

$$\text{Cl } M = \bigcap_D (D \circ M \circ D),$$

where D runs through the symmetric entourages of X.

For let D be a symmetric entourage. Then $(x, y) \in D \circ M \circ D$ if and only if there exists a point $(\xi, \eta) \in M$ such that $(x, \xi) \in D$ and $(\eta, y) \in D$, in other words, such that $x \in D[\xi]$ and $y \in D[\eta]$, or again such that

$$(x, y) \in D[\xi] \times D[\eta].$$

Since $D[\xi] \times D[\eta]$ is a neighbourhood of (ξ, η) in the toplogical product $X \times X$ this proves the first assertion.

Moreover, because of the symmetry of D the same condition can also be written in the form

$$(\xi, \eta) \in D[x] \times D[y].$$

Now as D runs through the symmetric entourages of X the sets $D[x] \times D[y]$ form a neighbourhood base at the point (x, y) in $X \times X$. In fact, if D', D'' are any two entourages there exists a symmetric entourage D such that $D \subset D' \cap D''$ and then

$$D[x] \times D[y] \subset D'[x] \times D''[y].$$

Therefore, to say that $D[x] \times D[y]$ intersects M, for all symmetric entourages D, is to say that (x, y) is an adherence point of M. This completes the proof.

Corollary 8.12

If H is a subset of the uniform space X then

$$\text{Cl } H = \bigcap_D D[H],$$

where D runs through the symmetric entourages of X.

This follows by taking $M = H \times H$ in Proposition 8.11 and observing that

$$D \circ M \circ D = D[H] \times D[H],$$

in this case. For topological groups this relation was established earlier in Chapter 2.

Corollary 8.13

Let X be a uniform space. Then with respect to the product topology on $X \times X$,

(i) *the interiors of the entourages of X form a base for the uniformity, and*
(ii) *the closures of the entourages of X form a base for the uniformity.*

For if D is any entourage of X there exists a symmetric entourage D' such that $(D')^3 \subset D$. Since $(D')^3$ is a neighbourhood of D', by Proposition 8.11, the interior of D in $X \times X$ contains D' and is therefore an entourage of X. This proves (i). Moreover,

$$D' \subset \mathrm{Cl}\, D' \subset (D')^3 \subset D,$$

and so D also contains the closure of an entourage of X, whence (ii).

Corollary 8.14

Let X be a uniform space. Then X is regular, in the uniform topology

For let $D[\xi]$ be a uniform neighbourhood of the point ξ of X, where D is an entourage of X. Choose an entourage D' of X such that $D' \circ D' \subset D$. Then $D'[\xi]$ is a uniform neighbourhood of ξ such that $\mathrm{Cl}(D'[\xi]) \subset D[\xi]$, which establishes regularity. Of course, we have already shown this, in Proposition 6.19, for the case when X is a topological group.

We see from Corollary 8.14 that no non-regular topological space can be given uniform structure in a way which is compatible with the topology. For example, in the case of an infinite set the cofinite topology cannot arise from a uniform structure.

Returning to the general theory we prove

Proposition 8.15

Let A be a dense subset of the uniform space X, in the uniform topology. Then the closures in $X \times X$ of the entourages of A, in the relative uniformity, form a base for the entourages of X.

For $X \times A$ is dense in $X \times X$ since A is dense in X. So let E be an open entourage of A. Then E is the trace on $A \times A$ of an open entourage D of X.

However, D is contained in the closure of E which in turn is contained in the closure of D. This proves Proposition 8.15.

As we have seen in the previous chapter there is a separated quotient space X' associated with each uniform space X. Since X is regular, in the uniform topology, there is also a Hausdorff quotient space associated with X, according to the procedure described in Chapter 6. It makes a good exercise to check that this Hausdorff space is just X', with the uniform topology.

Next we come to a series of results which involve compactness in various ways.

Proposition 8.16

Let Γ be an open covering of the compact uniform space X. Then there exists an entourage D such that each of the uniform neighbourhoods $D[x]$ $(x \in X)$ is contained in some member of Γ.

For given x there exists a uniform neighbourhood $D_x[x]$ of x which is contained in some member of Γ. Let E_x be an entourage such that $E_x \circ E_x \subset D_x$. The family of neighbourhoods $E_x[x]$ $(x \in X)$ covers the compact X. Extract a finite subcovering indexed by $x_1, \ldots, x_n$, say. Then the intersection $E_{x_1} \cap \cdots \cap E_{x_n}$ is an entourage with the required property.

Proposition 8.17

Let $\phi \colon X \to Y$ be continuous, where X and Y are uniform spaces. Suppose that X is compact. Then ϕ is uniformly continuous.

Let E be any entourage of Y, and let F be a symmetric entourage of Y such that $F \circ F \subset E$. Since ϕ is continuous there exists, for each point x of X, an open neighbourhood $V_x \subset \phi^{-1}F[\phi(x)]$. The family $\{V_x \colon x \in X\}$ constitutes an open covering of the compact X and so, by Proposition 8.16, there exists an entourage D of X such that each uniform neighbourhood $D[\xi]$, where $\xi \in X$, is contained in V_x for some x. Thus $D[\xi] \subset \phi^{-1}F[\phi(x)]$, in particular $\phi(\xi) \in F[\phi(x)]$ and so $\phi(x) \in F[\phi(\xi)]$, therefore

$$\phi D[\xi] \subset F \circ F[\phi(\xi)] \subset E[\phi(\xi)].$$

Since this is true for all ξ we obtain that $(\phi \times \phi)D \subset E$, and so ϕ is uniformly continuous, as asserted.

Proposition 8.18

Let H, K be disjoint subspaces of the uniform space X, with H compact and K closed. Then there exists an entourage D of X such that the uniform neighbourhoods D[H] and D[K] are disjoint.

For, if not, the trace on H of the family of sets $D \circ D[K]$, where D runs through the symmetric entourages of X, is non-empty and consists of non-empty sets, and so the trace generates a filter on H. By compactness this filter has an adherence point x_0 say, where $x_0 \in H$. Then for each symmetric entourage D of X the uniform neighbourhood $D^3[x_0]$ of x_0 intersects K and so $x_0 \in K$, since K is closed. Thus we have a contradiction which establishes Proposition 8.18.

Corollary 8.19

Let H be a compact subspace of the uniform space X. The uniform neighbourhoods D[H], as D runs through the entourages of X, form a base for the neighbourhoods of H.

For let U be an open neighbourhood of H. The complement $K = X - U$ is closed and disjoint from H. So there exists, by Proposition 8.16, an entourage D of X such that $D[H]$ does not intersect $D[K]$. Then $D[H] \subset U$, as required.

8.3 Uniformization of compact Hausdorff spaces

Proposition 8.20

Let X be a compact Hausdorff space. Then the neighbourhood filter of the diagonal ΔX in $X \times X$ constitutes a uniformity on X such that the uniform topology coincides with the original topology. Moreover, no other uniformity on X has this property.

First, we have to show that the family $\mathcal{N}$ of neighbourhoods of ΔX satisfies the conditions for a uniformity and that the uniform topology is the original topology. For this purpose it is sufficient to show that $\mathcal{N}$ forms a separated uniformity since in that case the uniform topology cannot be finer than the original topology and so is identical to it, from Proposition 6.15.

Note that the intersection of the members of $\mathcal{N}$ is precisely ΔX so that $\mathcal{N}$, if a uniformity, is separated. For if $\xi \neq \eta$ in X the complement of the point (ξ, η) in $X \times X$ is a neighbourhood of ΔX, i.e. a member of $\mathcal{N}$.

Clearly $\mathcal{N}$ satisfies all the axioms for a uniformity on X except, possibly, for the last. Suppose, to obtain a contradiction, that $\mathcal{N}$ does not satisfy the condition of weak transitivity. Then there exists a member N of $\mathcal{N}$ such that for all members M of $\mathcal{N}$ the set $M^2 \cap (X^2 - N)$ is non-empty. The sets $M^2 \cap (X^2 - N)$, where M runs through the members of $\mathcal{N}$, form a base for a filter on X^2, and this filter has an adherence point (ξ, η), say, by compactness. Note that (ξ, η) is not in ΔX, since $X^2 - N$ does not intersect ΔX.

Now X is Hausdorff, by hypothesis, and so there exist open neighbourhoods U of ξ and V of η which are disjoint. Also X is regular, by Corollary 8.14, and so there exist closed neighbourhoods $U' \subset U$ and $V' \subset V$ of ξ and η, respectively. Write $W = X - (U' \cup V')$ and consider the neighbourhood

$$K = (U \times U) \cup (V \times V) \cup (W \times W)$$

of ΔX in $X \times X$. The definitions imply that $(U' \times X) \cap K = U' \times U$ and $(U \times X) \cap K = U \times (X - V)$. Hence the neighbourhood $U' \times V'$ of (ξ, η) in $X \times X$ does not intersect $K \circ K$. Thus we have obtained a contradiction and established the first part of Proposition 8.20.

To demonstrate the second part, suppose given a uniformity on X, compatible with the topology. Then the entourages are neighbourhoods of ΔX, as we have seen, but the converse may be false for non-compact X, as Figure 8.2 illustrates.

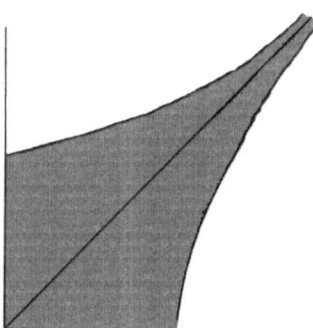

Figure 8.2. Neighbourhood of diagonal but not an entourage.

Suppose, to obtain a contradiction, that there exists a neighbourhood W of ΔX which is not an entourage.

Without real loss of generality we may take W to be open. The entourage filter of X traces a filter $\mathscr{F}$ on the complement CW of W. Now CW is closed in the compact space $X \times X$ and so is compact. Therefore $\mathscr{F}$ admits an adherence point (x, y), say, in CW. The closed entourages form a base, as we have seen, and so their intersection is ΔX, since X is Hausdorff. Therefore $(x, y) \in \Delta X$, since (x, y) belongs to every closed entourage. But $(x, y) \in CW$ which is disjoint from ΔX and so we have our contradiction.

8.4 Cauchy sequences

The reader is doubtless already familiar with the concept of Cauchy convergence, in relation to sequences of real numbers. Essentially, the same definition can be used in relation to sequences in any metric space or, more generally, in any uniform space.

Definition 8.21

Let $\langle x_n \rangle$ be a sequence of points of the uniform space X. The sequence satisfies the Cauchy condition if for each entourage D of X there exists an integer k such that $(x_n, x_m) \in D$ whenever $n, m \geq k$.

When the condition in Definition 8.21 is satisfied we describe $\langle x_n \rangle$ as a *Cauchy sequence*. Note that it is sufficient if the condition is satisfied for all members of a base for the uniformity. For example, when X is metric it is sufficient if the condition is satisfied for all ε-entourages $U_\varepsilon (\varepsilon > 0)$.

Proposition 8.22

Let $\langle x_n \rangle$ be a sequence of points of the uniform space X. If $\langle x_n \rangle$ converges in the uniform topology then $\langle x_n \rangle$ is a Cauchy sequence.

For let x be a limit point of $\langle x_n \rangle$. Given an entourage D of X choose a symmetric entourage D' of X such that $D' \circ D' \subset D$. Then $D'[x]$ is a uniform neighbourhood of x and so there exists an integer k such that $x_n \in D'[x]$ whenever $n \geq k$. So if $m, n \geq k$ we have $x_m, x_n \in D'[x]$, and consequently $(x_m, x_n) \in D' \circ D' \subset D$. Therefore $\langle x_n \rangle$ satisfies the Cauchy condition.

For an example of a Cauchy sequence which is not convergent take the sequence $\langle 1/n \rangle$ in the space $\mathbb{R}_+$ of positive real numbers, with the euclidean uniformity.

Proposition 8.23

Let $\phi\colon X \to Y$ be a uniformly continuous function, where X and Y are uniform spaces. If $\langle x_n \rangle$ is a Cauchy sequence in X then $\langle \phi(x_n) \rangle$ is a Cauchy sequence in Y.

For let E be an entourage of Y. Then $D = (\phi \times \phi)^{-1}E$ is an entourage of X. If $\langle x_n \rangle$ satisfies the Cauchy condition in X there exists an integer k such that $(x_n, x_m) \in D$ whenever $n, m \geq k$. Then $(\phi(x_n), \phi(x_m)) \in E$, for $n, m \geq k$, and so $\langle \phi(x_n) \rangle$ satisfies the Cauchy condition in Y.

It should be remarked that a function $\phi\colon X \to Y$, where X and Y are uniform spaces, may transform Cauchy sequences into Cauchy sequences and yet fail to be uniformly continuous. The exponential function $\mathbb{R} \to \mathbb{R}$, where $\mathbb{R}$ has the euclidean uniformity, provides an example of this.

In the case of a metric space the condition of total boundedness can be expressed in terms of the Cauchy condition as follows.

Proposition 8.24

The metric space X is totally bounded if and only if each sequence in X contains a Cauchy subsequence.

For suppose that X is totally bounded. Let $\langle x_n \rangle$ be a sequence of points of X. Since X admits a finite ε-net for $\varepsilon = 1/2$ we can cover X by a finite number of open balls of radius $1/2$. At least one of these balls must contain a subsequence $\langle x_n^{(1)} \rangle$ say of the original sequence, and the distance between any two terms of this subsequence is always less than 1. Similarly, with $\varepsilon = 1/4$ we can obtain a subsequence $\langle x_n^{(2)} \rangle$ of $\langle x_n^{(1)} \rangle$ in which the distance between any two terms is less than $1/2$. Proceeding by induction, for $\varepsilon = 1/2k$ we obtain a subsequence $\langle x_n^{(k)} \rangle$ of $\langle x_n^{(k-1)} \rangle$ in which the distance between any two terms is less than $1/k$, for $k = 2, 3, \ldots$.

The diagonal sequence $\langle x_n^{(n)} \rangle$ is a Cauchy sequence: for if $\varepsilon > 0$ is given we have only to take $k > 1/\varepsilon$ and then $\rho(x_m^{(m)}, x_n^{(n)}) < \varepsilon$ whenever $m, n \geq k$. Since $\langle x_n^{(n)} \rangle$ is a subsequence of the original sequence $\langle x_n \rangle$ this proves Proposition 8.24 in one direction.

For the converse, suppose that X is not totally bounded. Then there exists an $\varepsilon > 0$ such that X admits no finite ε-net. By the following construction we can obtain a sequence $\langle x_n \rangle$ such that $\rho(x_i, x_j) \geq \varepsilon$ whenever $i \neq j$; clearly, such a sequence is not only not a Cauchy sequence itself but cannot contain a Cauchy subsequence.

To construct the sequence, take x_1 to be any point of X. Since the finite set $\{x_1\}$ is not an ε-net for X there exists a point $x_2 \notin U_\varepsilon(x_1)$, i.e. a point such that $\rho(x_1, x_2) \geq \varepsilon$. Now the finite set $\{x_1, x_2\}$ is also not an ε-net for X, and so there exists a point $x_3 \notin U_\varepsilon(x_1) \cup U_\varepsilon(x_2)$, i.e. a point x_3 such that $\rho(x_1, x_3) \geq \varepsilon$ and $\rho(x_2, x_3) \geq \varepsilon$. We now proceed by induction.

If we have defined a set of points $\{x_1, x_2, \ldots, x_n\}$ such that $\rho(x_i, x_j) \geq \varepsilon$ whenever $i \neq j$ then this finite set is not an ε-net for X, and so there exists a point

$$x_{n+1} \notin U_\varepsilon(x_1) \cup U_\varepsilon(x_2) \cup \cdots \cup U_\varepsilon(x_n),$$

i.e. such that $\rho(x_i, x_j) \geq \varepsilon$ whenever $i \neq j$ for $i, j \leq n + 1$.

8.5 Cauchy filters

We have seen in Chapter 6 that for topological spaces in general the notion of convergent filter is more satisfactory than the notion of convergent sequence. This suggests the introduction of a Cauchy condition for filters on a uniform space, as follows.

Definition 8.25

Let $\mathcal{F}$ be a filter on the uniform space X. The filter satisfies the Cauchy condition if for each entourage D of X there exists a member M of $\mathcal{F}$ which is D-small in the sense that $M \times M \subset D$.

When the condition in Definition 8.25 is satisfied we describe $\mathcal{F}$ as a *Cauchy filter*. Note that it is sufficient if the condition is satisfied for all members of a base for the uniformity.

In the case of the discrete uniformity only the principal filters are Cauchy. In the case of the trivial uniformity every filter is Cauchy.

Clearly, if $\mathcal{F}$ is the elementary filter associated with a sequence $\langle x_n \rangle$ of points of the uniform space X then $\mathcal{F}$ is a Cauchy filter if and only if $\langle x_n \rangle$ is a Cauchy sequence.

The analogue of Proposition 8.22 for filters is

Proposition 8.26

Let $\mathcal{F}$ be a filter on the uniform space X. If $\mathcal{F}$ converges in the uniform topology then $\mathcal{F}$ is a Cauchy filter.

First, observe that every refinement of a Cauchy filter also satisfies the Cauchy condition. If $\mathscr{F}$ converges to the point x of X then $\mathscr{F}$ is a refinement of the neighbourhood filter $\mathscr{N}_x$. So the result will follow once we have proved that $\mathscr{N}_x$ is a Cauchy filter. But given an entourage D of X there exists an entourage D' of X such that $D' \circ D' \subset D$, and then the uniform neighbourhood $D'[x]$ of x is D-small. This proves Proposition 8.26. Of course, Proposition 8.22 can be obtained by applying this result to the case of an elementary filter.

Proposition 8.27

Let $\mathscr{F}$ be a Cauchy filter on the uniform space X. Then each adherence point of $\mathscr{F}$ is also a limit point of $\mathscr{F}$.

First, observe that for each entourage D of X there exists a closed member of $\mathscr{F}$ which is D-small. For since the closed entourages form a base, as we have seen, there exists a closed entourage E of X contained in D; choose M to be E-small, then $M \times M \subset E$ and so $N \times N \subset E \subset D$, where $N = \mathrm{Cl}\, M$. Also N is a member of $\mathscr{F}$, since $M \subset N$, and so N is the closed member, as required.

Now let x be an adherence point of $\mathscr{F}$. Then $x \in N$, since N is closed, and so $N \subset D[x]$, since N is D-small. Since the uniform neighbourhoods $D[x]$ of x form a base for $\mathscr{N}_x$ this shows that $\mathscr{F}$ is a refinement of $\mathscr{N}_x$, i.e. that $\mathscr{F}$ converges to x, as asserted.

Corollary 8.28

Let $\langle x_n \rangle$ be a Cauchy sequence in the uniform space X. If a subsequence of $\langle x_n \rangle$ converges to the point x of X then $\langle x_n \rangle$ converges to x.

For let $\mathscr{F}$ be the elementary filter associated with $\langle x_n \rangle$, and let $\mathscr{G}$ be the elementary filter associated with the subsequence, which is a refinement of $\mathscr{F}$. Then $\mathscr{G}$ converges to x, since the subsequence converges to x. Thus x is an adherence point of $\mathscr{G}$ and so is an adherence point of $\mathscr{F}$. Therefore x is a limit point of $\mathscr{F}$, by Proposition 8.27, since $\mathscr{F}$ is Cauchy, and so x is a limit point of $\langle x_n \rangle$, as asserted.

Proposition 8.29

Let $\phi \colon X \to Y$ be a uniformly continuous function, where X and Y are uniform spaces. If $\mathscr{F}$ is a Cauchy filter on X then $\phi_* \mathscr{F}$ is a Cauchy filter on Y.

For let E be an entourage of Y. Then the inverse image $D = (\phi \times \phi)^{-1}E$ is an entourage of X. Since $\mathscr{F}$ is Cauchy there exists a member M of $\mathscr{F}$ which is D-small. The direct image ϕM is a member of $\phi_*\mathscr{F}$ which is E-small. Therefore $\phi_*\mathscr{F}$ is Cauchy, as asserted.

Proposition 8.30

Let $\phi: X \to Y$ be a function, where X and Y are uniform spaces. Suppose that the uniformity of X is induced by ϕ from the uniformity of Y. If $\mathscr{G}$ is a Cauchy filter on Y such that $\phi^*\mathscr{G}$ is defined then $\phi^*\mathscr{G}$ is Cauchy as a filter on X.

For let E be an entourage of Y. If N is an E-small subset of Y then $\phi^{-1}N$ is a D-small subset of X, where $D = (\phi \times \phi)^{-1}E$. Since X has the induced uniformity the inverse images $(\phi \times \phi)^{-1}E$, as E runs through the entourages of Y, form a base for the entourages of X. Now Proposition 8.30 follows at once.

The notion of totally bounded uniform space was introduced in the previous chapter. An alternative characterization is provided by

Proposition 8.31

The uniform space X is totally bounded if and only if each filter $\mathscr{F}$ on X admits a Cauchy refinement.

For let X be totally bounded. Given an entourage D of X choose a symmetric entourage D' of X such that $D' \circ D' \subset D$. Since X is totally bounded there exists a finite subset S of X such that $D'[S] = X$. Each of the sets $D'[x]$ $(x \in S)$ is D-small and together they cover X. Given a filter $\mathscr{F}$ on X let $\mathscr{F}'$ be an ultrafilter refining $\mathscr{F}$. Since $X \in \mathscr{F}'$ some member $D'[x]$ of the covering is a member of $\mathscr{F}'$, by Proposition 1.18. Therefore $\mathscr{F}'$ is Cauchy. (It is only in this part of the argument that ultrafilters are used.)

For the converse, suppose that X is not totally bounded. Then there exists an entourage D of X such that $D[S]$ is a proper subset of X for every finite subset S of X. The family

$$\Gamma = \{X - D[S]\},$$

where S runs through the finite subsets of X, is a base for a filter $\mathscr{F}$ on X. I assert that $\mathscr{F}$ does not admit a Cauchy refinement.

For suppose, to obtain a contradiction, that there exists a Cauchy refinement $\mathscr{F}'$ of $\mathscr{F}$. Choose a D-small member M of $\mathscr{F}'$. Then M intersects $X - D[S]$, for each finite subset S of X, since $X - D[S]$ is a member of $\mathscr{F}$ and hence of $\mathscr{F}'$. Choose such an S and let x be a point of $M \cap (X - D[S])$. Then $M \subset D[x]$

and so M does not intersect $X - D[S']$, where $S' = S \cup \{x\}$. But $X - D[S']$ is also a member of $\mathscr{F}$, since S' is finite, and so $X - D[S']$ is a member of $\mathscr{F}'$. Thus we have our contradiction and the proof of Proposition 8.31 is complete.

Corollary 8.32

The uniform space X is totally bounded if and only if each ultrafilter on X is a Cauchy filter.

8.6 Uniformization of function spaces

Consider the set Y^X of functions $X \to Y$, where X is a set and Y is a uniform space. In this situation the product uniformity on Y^X is usually known as the *uniformity of pointwise convergence*. The associated uniform topology is the topology of pointwise convergence already discussed in Chapter 5.

In fact, the uniformity of pointwise convergence is generally less important than a certain refinement, called the *uniformity of uniform convergence*, which is defined as follows. Consider for each entourage D of Y the subset $\tilde{D}$ of $Y^X \times Y^X$ consisting of pairs (θ, ϕ) of functions $\theta, \phi \colon X \to Y$ such that

$$(\theta(x), \phi(x)) \in D \qquad (x \in X).$$

The family of subsets $\tilde{D}$, as D runs through the entourages of Y, constitutes a base for the uniformity of uniform convergence. Figure 8.3 suggests the uniformities of uniform convergence for real-valued functions.

When dealing with Y^X, and with subsets of Y^X, it is essential to be clear as to which uniformity and associated topology is being used. Thus a sequence or filter

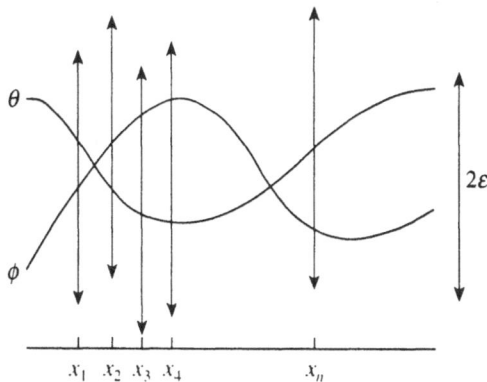

Figure 8.3.

may be pointwise Cauchy but not uniformly Cauchy, or may be pointwise convergent but not uniformly convergent.

Note that if Y is separated then so is Y^X with either uniformity. For Y^X is separated in the uniformity of pointwise convergence, by Proposition 7.15 and so is separated in any finer uniformity.

Obviously, a uniformly convergent filter is uniformly Cauchy and pointwise convergent, and similarly with sequences. In the other direction we have

Proposition 8.33

Let Φ be a family of functions $X \to Y$, where X is a set and Y is a uniform space. Let $\mathscr{F}$ be a uniformly Cauchy filter on Φ which converges pointwise to some member ϕ of Φ. Then $\mathscr{F}$ converges uniformly to ϕ.

Corollary 8.34

Let $\langle \phi_n \rangle$ be a uniformly Cauchy sequence in Φ, as above, which converges pointwise to ϕ. Then $\langle \phi_n \rangle$ converges uniformly to ϕ.

The corollary follows by applying the proposition to the elementary filter $\mathscr{F}$ associated with the sequence. To prove the proposition itself, let D be any entourage of Y. There exists a symmetric entourage E of Y such that $E \circ E \subset D$. Since $\mathscr{F}$ is uniformly Cauchy there exists an $\tilde{E}$-small member M of $\mathscr{F}$. Let $\theta \in M$ and let $x \in X$. Then $\pi_{x*}\mathscr{F}$ converges to $\pi_x(\phi)$ in Y, and so the uniform neighbourhood $E[\pi_x(\phi)]$ of $\pi_x(\phi)$ intersects $\pi_x M$ in $\pi_x(\xi)$, say, where $\xi \in M$. Now $(\pi_x(\theta), \pi_x(\xi)) \in E$, since $\theta, \xi \in M$ and M is $\tilde{E}$-small, while $(\pi_x(\xi), \pi_x(\phi)) \in E$ by choice of ξ. Therefore

$$(\pi_x(\theta), \pi_x(\phi)) \in E \circ E \subset D.$$

In other words, $(\theta, \phi) \in \tilde{D}$, i.e. $\theta \in \tilde{D}[\phi]$. Since this is true for all $\theta \in M$ we obtain $M \subset \tilde{D}[\phi]$ and hence $\tilde{D}[\phi] \in \mathscr{F}$. Therefore $\mathscr{F}$ converges uniformly to ϕ, as asserted.

Function-spaces provide some interesting illustrations. Thus consider the group G of uniform equivalences $\theta \colon T \to T$, where T is a uniform space, with the uniformity of uniform convergence. In the associated topology, i.e. the topology of uniform convergence, a base for the neighbourhood filter of the neutral element consists of the subsets W_D, for all entourages D of T, where

$$W_D = \{\theta \colon (\theta(t), t) \in D \, \forall t \in T\}.$$

The right relation determined by W_D consists of pairs (ϕ, ψ) of elements of G such that $\psi\phi^{-1} \in W_D$, i.e. such that $(\psi(s), \phi(s)) \in D$ for all s. Thus the right uniformity is just the uniformity of uniform convergence again but in general the left uniformity is not. The same situation arises if we form G from the group of topological self-equivalences rather than uniform self-equivalences.

For another illustration consider the group G of bijections $\theta: T \to T$, where T is an infinite set. Regarding T as a discrete uniform space let us give G the uniformity of pointwise convergence. In the associated topology, i.e. the topology of pointwise convergence, a base for the neighbourhood filter of the neutral element consists of the subsets

$$W_S = \{\theta: \theta(t) = t \, \forall t \in S\},$$

where S runs through the finite subsets of T. The left relation determined by W_S consists of pairs (ϕ, ψ) of elements of G such that $\phi^{-1}\psi(t) = t$ for all $t \in S$, i.e. such that $\psi(t) = \phi(t)$ for all $t \in S$. Thus the left uniformity is just the uniformity of pointwise convergence.

In the case of a topological group a sequence of elements may satisfy the Cauchy condition in the right uniformity but not in the left. For example consider the group G of homeomorphisms $\phi: I \to I$, where $I = [0, 1]$ denotes the closed unit interval with the relative uniform structure obtained from $\mathbb{R}$. As we have already seen, G is a topological group, with respect to the topology of uniform convergence, and the right uniformity coincides with the uniformity of uniform convergence. Consider the sequence $\langle \phi_n \rangle$ in G which is given for $n = 2, 3, \ldots$, by

$$\phi_n(t) = \begin{cases} \frac{2t}{n} & (0 \le t \le \frac{1}{2}) \\ -1 + \frac{2}{n} + t(2 - \frac{2}{n}) & (\frac{1}{2} \le t \le 1) \end{cases}$$

Let $\varepsilon > 0$, $k \ge 4/\varepsilon$ and $n \ge k$; then for $0 \le t \le \frac{1}{2}$ and $p = 1, 2, \ldots$, we have $|\phi_{n+p}(t) - \phi_n(t)| = \frac{2tp}{n(n+p)} \le \frac{2}{k} \le \varepsilon$; while for $\frac{1}{2} \le t \le 1$ we have

$$|\phi_{n+p}(t) - \phi_n(t)| \le \frac{2p}{n(n+p)} + \frac{2tp}{n(n+p)} \le \frac{4}{k} \le \varepsilon.$$

Thus $\langle \phi_n \rangle$ is a Cauchy sequence with the right uniformity. However, since

$$\phi_{2n}^{-1}\left(\frac{1}{2n}\right) - \phi_n^{-1}\left(\frac{1}{2n}\right) = \frac{1}{2} - \frac{1}{4} = \frac{1}{4},$$

for all $n \ge 2$, we see that $\langle \phi_n \rangle$ is not a Cauchy sequence in the left uniformity, in fact $\langle \phi_n \rangle$ does not converge in G. Another example with similar properties is the group G of bijections $\phi: T \to T$, where T is an infinite discrete space and G is given the topology of pointwise convergence. It should be noted that both these examples satisfy the Hausdorff condition.

Exercises

8.1. In the uniformity on the set X determined by finite partitions, as on page 128, show that every ultrafilter is Cauchy and deduce that for infinite X the uniformity is not discrete.

8.2. Let $\phi\colon \mathbb{R} \to \mathbb{R}$ be given by $\phi(\xi) = \xi^3$. Give the codomain $\mathbb{R}$ the euclidean uniformity and give the domain $\mathbb{R}$ the uniformity induced by ϕ. Show that the latter uniformity is a strict refinement of the former but that the Cauchy sequences are the same in both cases.

8.3. Let $\mathscr{F}$ be a non-principal ultrafilter on the infinite set X. Show that the family of subsets of $X \times X$ consisting of the unions of the diagonal ΔX with the products $M \times M$, where M runs through the members of $\mathscr{F}$, is a base for a uniformity on X and that the associated uniform topology is discrete.

8.4. Let Γ be an open covering of the compact Hausdorff space X. Show that there exists a neighbourhood D of the diagonal of X such that $D[x]$ is contained in some member of Γ for each point x of X.

8.5. Let D be a closed entourage of the uniform space X. Show that $D[A]$ is closed for each compact subset A of X.

8.6. Let D be an entourage of the uniform space X. Show that for each subset A of X the union of the sets $D^n[A]$, for all positive integers n, is both open and closed in X.

8.7. Let H be a subgroup of the topological group G. Suppose that the left and right uniformities of H coincide. Show that the same is true of the closure $\mathrm{Cl}\, H$ of H.

8.8. Let G be the group of all bijections $\phi\colon \mathbb{Z} \to \mathbb{Z}$ such that $\phi(n) = n$ for all but a finite number of integers n. Give G the topology of pointwise convergence, with respect to the discrete topology of $\mathbb{Z}$. Show that the left and right uniformities of G are different.

9
Connectedness

9.1 Connected spaces

Definition 9.1

The topological space X is connected if for each discrete $\mathbb{D}$ every continuous function $\lambda\colon X \to \mathbb{D}$ is constant.

For example, X is connected if X has the trivial topology, since if d is a value of λ then $\lambda^{-1}(d)$ is closed and non-empty, therefore full. On the other hand, X is not connected if X has the discrete topology and more than one point, since in that case we can take $X = \mathbb{D}$ and λ the identity. Again, the punctured real line $\mathbb{R}_* = \mathbb{R} - \{0\}$ is not connected, since we can take $\mathbb{D} = \{-1, +1\} \subset \mathbb{R}$ and define λ by $\lambda(x) = -1$ for $x < 0$, $\lambda(x) = +1$ for $x > 0$.

Proposition 9.2

The topological space X is connected if and only if X contains no open and closed subset, other than the empty set and the full set.

For suppose that $\lambda\colon X \to \mathbb{D}$ is continuous, where $\mathbb{D}$ is discrete. Then $\lambda^{-1}(d)$ is open and closed in X, for any point d of $\mathbb{D}$. If d is a value of λ then $\lambda^{-1}(d)$ is non-empty, while if λ is non-constant then $\lambda^{-1}(d)$ is non-full. This proves Proposition 9.2 in one direction.

For the proof in the other direction, suppose that there exists an open and closed subset H of X which is neither empty nor full. Choose $\mathbb{D} = \{-1, +1\} \subset \mathbb{R}$, with discrete topology. Define $\lambda\colon X \to \mathbb{D}$ by $\lambda(x) = +1$ if $x \in H$, $\lambda(x) = -1$ if $x \notin H$. Then λ is continuous and non-constant.

The characterization of connectedness provided by Proposition 9.2 is often convenient in practice. For example, it enables one to see at once, that a non-connected cofinite space must be the union of two finite sets and therefore finite. However, the actual definition (Definition 9.1) is generally more convenient when it comes to developing the theory.

Proposition 9.3

The real line $\mathbb{R}$ is connected, with the euclidean topology.

For let $\lambda\colon \mathbb{R} \to \mathbb{D}$ be continuous, where $\mathbb{D}$ is discrete. Let d be any point in the image of λ: to fix ideas let $d = \lambda(0)$. Suppose, to obtain a contradiction, that $\lambda(t) \neq d$ for some $t \in \mathbb{R}$. Then $t \neq 0$. In the argument which follows we assume $t > 0$; if $t < 0$ we use the same argument with λ replaced by $-\lambda$.

Consider the closed interval $J = [0, t] \subset \mathbb{R}$, which is bounded and so compact. We have $J = H \cup K$, where

$$H = \{\xi \in J \colon \lambda(\xi) = d\}, \qquad K = \{\eta \in J \colon \lambda(\eta) \neq d\}.$$

Of course, H and K are disjoint, and non-empty since $0 \in H$ and $t \in K$. Also H and K are open and closed, since $\lambda|J$ is continuous. Since H and K are closed and bounded they are compact. Therefore $H \times K$ is compact and so the continuous function $\rho\colon H \times K \to \mathbb{R}$ attains its infimum, where

$$\rho(\xi, \eta) = |\xi - \eta| \qquad (\xi \in H, \eta \in K).$$

Let $\alpha \in H$, $\beta \in K$ be such that $|\alpha - \beta|$ is the infimum of ρ. Then $\gamma = \frac{1}{2}(\alpha + \beta) \in J$, since $\alpha, \beta \in J$ and J is an interval. But $\gamma \notin H$, since $\rho(\gamma, \beta) = \frac{1}{2}|\alpha - \beta|$, and $\gamma \notin K$, since $\rho(\alpha, \gamma) = \frac{1}{2}|\alpha - \beta|$. Since $J = H \cup K$ we have our contradiction.

Before discussing the consequences of Proposition 9.3 it is illuminating to examine the situation for the rational line $\mathbb{Q}$, which is a complete contrast. A topological space which is not connected is often said to be disconnected. However, this should not be confused with the following much stronger condition:

Definition 9.4

The topological space X is totally disconnected if all the connected subspaces of X are one-point sets.

Discrete spaces are totally disconnected, obviously, but these are not the only spaces to satisfy the condition. For example, consider the rational line $\mathbb{Q}$, with the euclidean topology. Let H be any subset of $\mathbb{Q}$ with at least two points α, β, say, where $\alpha < \beta$. Choose any irrational ξ such that $\alpha < \xi < \beta$, and consider the function $\lambda \colon H \to \{-1, +1\}$ given by $\lambda(x) = -1$ for $x < \xi$, $\lambda(x) = +1$ for $x > \xi$. Then λ is continuous and non-constant, so that H is disconnected. Therefore $\mathbb{Q}$ is totally disconnected.

Proposition 9.5

Let $\phi \colon X \to Y$ be a continuous surjection, where X and Y are topological spaces. If X is connected then so is Y.

For let $\lambda \colon Y \to \mathbb{D}$ be continuous, where $\mathbb{D}$ is discrete. Then $\lambda\phi \colon X \to \mathbb{D}$ is continuous. If X is connected then $\lambda\phi$ is constant and so λ is constant. It follows at once from Proposition 9.5 that connectedness is a topologically invariant property.

Proposition 9.6

Let H be a connected subspace of the topological space X. Then the closure Cl H is also connected.

For let $\lambda \colon \text{Cl } H \to \mathbb{D}$ be continuous, where $\mathbb{D}$ is discrete. Since H is connected and $\lambda|H$ is continuous then λ is constant on H with value d, say. Now $\lambda|H = \delta|H$, where $\delta \colon \text{Cl } H \to \mathbb{D}$ is constant at d. By Proposition 6.9 the coincidence set of λ and δ is closed in Cl H. But the coincidence set contains H and so contains Cl H. Therefore $\lambda = \delta$ throughout Cl H and so λ is constant. We conclude that Cl H is connected, as asserted. One also obtains that if $H \subset K \subset \text{Cl } H$ then K is connected, by applying Proposition 9.6 to Cl H in place of X.

We have seen in Proposition 9.3 that the real line $\mathbb{R}$ is connected, with the euclidean topology. Since connectedness is a topologically invariant property, we conclude that any open interval (α, β), say, is also connected and hence, using Proposition 9.6, that any closed interval $[\alpha, \beta]$ is also connected. Similarly, we conclude that half-open intervals such as $(\alpha, \beta]$ are also connected, since (α, β) is dense in $(\alpha, \beta]$.

Open rays and closed rays are connected for similar reasons. Thus every interval in the real line $\mathbb{R}$ is connected; conversely, every non-empty connected subset of $\mathbb{R}$ is an interval. For let H be a non-empty subset. If H is not an interval then there exist $\alpha, \beta \in H$ and $\xi \in \mathbb{R} - H$ such that $\alpha < \xi < \beta$. Define $\lambda \colon H \to \mathbb{D}$, where $\mathbb{D} = \{-1, +1\} \subset \mathbb{R}$, by $\lambda(x) = -1$ if $x < \xi$, $\lambda(x) = +1$ if $x > \xi$. Then λ is continuous and non-constant, so that H cannot be connected. This leads at once to

Proposition 9.7

Let $\phi \colon X \to \mathbb{R}$ be a real-valued continuous function, where X is any connected topological space. If $\alpha, \beta \in \phi X$, where $\alpha < \beta$, then $\xi \in \phi X$ whenever $\alpha \leq \xi \leq \beta$.

This result, which is generally known as the *intermediate value theorem*, has some attractive applications, of which we give just two.

Corollary 9.8

Let $\phi \colon \mathbb{R} \to \mathbb{R}$ be a polynomial of odd degree. Then $\phi(\xi) = 0$ for some value of ξ.

Clearly $\phi(\xi)$ is positive and $\phi(-\xi)$ is negative for large enough ξ, say $\xi \geq \xi_0$. By Proposition 9.7, $\phi(\xi) = 0$ for some $\xi \in (-\xi_0, \xi_0)$, and so the result is obtained.

Corollary 9.9

Let $\phi \colon [0, 1] \to [0, 1]$ be a continuous function. Then $\phi(\xi) = \xi$ for some point $\xi \in [0, 1]$.

For suppose, to obtain a contradiction, that $\phi(\xi) \neq \xi$ for all $\xi \in [0, 1]$. In particular, $\phi(0) > 0$ and $\phi(1) < 1$. So the continuous function $\psi \colon [0, 1] \to \mathbb{R}$ defined by $\psi(\xi) = \phi(\xi) - \xi$ is positive when $\xi = 0$ and negative when $\xi = 1$, By Proposition 9.7, $\psi(\xi) = 0$ for some value of ξ, which gives us our contradiction.

Proposition 9.10

Let R be an equivalence relation on the topological space X for which each equivalence class is a connected subspace of X. If the quotient space X/R is also connected then X is connected.

For let $\lambda \colon X \to \mathbb{D}$ be continuous, where $\mathbb{D}$ is discrete. Since the equivalence classes are connected, λ is constant on each class, and so $\lambda = \mu\pi$, for some function $\mu\colon X/R \to \mathbb{D}$. Now μ is continuous, since X/R has the quotient topology, and so constant, since X/R is connected. Therefore λ is constant and so X is connected.

Corollary 9.11

Let G' be a subgroup of the topological group G. If G' and G/G' are connected then so is G.

Proposition 9.12

Let $\{A_j\}$ be a family of connected subspaces of the topological space X. Suppose that each member of the family intersects every other member of the family. Then the union A of the members of the family is connected.

For let $\lambda\colon A \to \mathbb{D}$ be continuous, where $\mathbb{D}$ is discrete. Since each of the A_j is connected and since $\lambda|A_j$ is continuous we have that $\lambda|A_j$ is constant with value ε_j, say. Since any two members of the family intersect the constant ε_j is independent of j. Thus λ is constant on A and so A is connected.

Corollary 9.13

Let $\{C_j\}$ be a family of connected subspaces of the topological space X. Let B be a connected subspace of X which intersects each of the C_j. Then $B \cup C$ is connected, where C denotes the union of the members of the family.

To see this, put $A_j = B \cup C_j$. Then Proposition 9.12 shows that A_j is connected. And then Proposition 9.12 again shows that $A = B \cup C$ is connected.

Proposition 9.14

Let $\{X_j\}$ be a family of connected spaces. Then the topological product $\prod X_j$ is also connected.

In the finite case this result can be proved in a straightforward fashion, using Proposition 9.12. In the general case, however, it is necessary to proceed in two steps, as follows.

The first step is to choose a point $\xi = (\xi_j)$ of $\prod X_j$ and consider the subset H of $\prod X_j$ consisting of points $x = (x_j)$ such that $x_j \neq \xi_j$ for at most a finite number of values of j. Then H is dense in $\prod X_j$. For if $\prod U_j$ is a restricted product of open sets then $\xi_j \notin U_j$ for at most a finite number of values of j and so $\prod U_j$ intersects H.

The second step is to show that H is connected and hence $\mathrm{Cl}\, H = \prod X_j$ is connected, by Proposition 9.6. To see this, let $\lambda \colon H \to \mathbb{D}$ be continuous, where $\mathbb{D}$ is discrete. Then $\rho_j \colon X_j \to \mathbb{D}$ is also continuous, where the value of ρ_j at the point x_j of X_j is given by evaluating λ at the point with the same coordinates as ξ except that ξ_j is replaced by x_j. Then ρ_j is constant, since X_j is connected, and moreover, the value of the constant is independent of j since $\rho_j(\xi_j) = \lambda(\xi)$ for all values of j. Therefore λ is constant throughout H. Thus H, and hence $\mathrm{Cl}\, H$, is connected, and the proof is complete.

Hence and from Proposition 9.3 we conclude that the real n-space $\mathbb{R}^n$ is connected, in the usual topology. It follows at once that the open n-ball U^n is connected and hence, using Proposition 9.6, that the closed n-ball B^n is connected. Similarly, the punctured n-sphere is connected. We deduce from this that the n-sphere S^n itself is connected when $n \geq 1$, since S^n is the union of the complement of the north pole and the complement of the south pole.

9.2 Connectedness components

We now come to the idea of connectedness component, which depends on

Definition 9.15

The points ξ, η of the topological space X are equivalent, in the sense of connectedness, if there exists a connected subspace C of X containing both ξ and η.

The relation defined in Definition 9.15 is obviously symmetric and reflexive; transitivity is an immediate consequence of Corollary 9.13. The equivalence classes are called the *connectedness components* of X, or simply *components* when this is unlikely to create confusion. The quotient space of X with respect to the equivalence relation is totally disconnected. When X itself is totally disconnected the components are the one-point subsets. It follows at once from Proposition 9.6 that components are closed subsets. Also we have

Proposition 9.16

For each point ξ of the topological space X the connectedness component containing ξ is the union of all the connected subspaces of X which contain ξ.

Clearly the union is contained in the connectedness component. Conversely, any point η in the component is contained in a connected subspace C of X which also contains ξ and so is contained in the union.

Obviously a topological space is connected if and only if it has precisely one connectedness component. If the number of components is finite then each is open as well as closed, since each is the complement of the union of the others. In general, however, the components are not open. For example, take the rational line $\mathbb{Q}$: the components are the one-point subsets.

We have seen in Proposition 9.3 that the real line $\mathbb{R}$, with the euclidean topology, is connected. We have also seen that the punctured line $\mathbb{R}_* = \mathbb{R} - \{0\}$ is disconnected. In fact $\mathbb{R}_*$ can be partitioned into the subspaces $(-\infty, 0)$ and $(0, +\infty)$, each of which is homeomorphic to $\mathbb{R}$ and therefore connected. Thus $\mathbb{R}_*$ has two components, one for the negative reals and one for the positive. More generally, the topological space obtained from $\mathbb{R}$ by removing n distinct points $(n = 0, 1, \ldots)$ has precisely $n + 1$ components.

We have also seen that the circle S^1 is connected. However, the punctured circle, being homeomorphic to $\mathbb{R}$, is still connected. More generally, the topological space obtained from S^1 by removing n distinct points $(n = 1, 2, \ldots)$ has precisely n components.

The number of connectedness components of a topological space is obviously a topological invariant; if two spaces do not have the same number of components they cannot be homeomorphic. This criterion can be used directly, of course, but it can also be used indirectly, as follows. Suppose that for *some* set $\{x_1, \ldots, x_n\}$ of n distinct points of the topological space X and for *every* set $\{y_1, \ldots, y_n\}$ of n distinct points of the topological space Y the number of components of $X - \{x_1, \ldots, x_n\}$ is different from the number of components of $Y - \{y_1, \ldots, y_n\}$. Then X and Y cannot be homeomorphic, since a homeomorphism $\phi \colon X \to Y$ would determine a homeomorphism

$$X - \{x_1, \ldots, x_n\} \to Y - \{y_1, \ldots, y_n\},$$

with $y_i = \phi(x_i)$ $(i = 1, \ldots, n)$.

For example, consider bounded intervals of the real line $\mathbb{R}$, with the euclidean topology. If we remove the end-point α from the closed interval $[\alpha, \beta]$, where $\alpha < \beta$, we obtain the half-closed interval $(\alpha, \beta]$, which is connected. But if we remove any point from (α, β) the resulting space is disconnected, since (α, β) is homeomorphic to $\mathbb{R}$. Therefore $[\alpha, \beta]$ is not homeomorphic to (α, β). However, to show that $(\alpha, \beta]$ is not homeomorphic to $[\alpha, \beta]$ a refinement of the argument is needed, as follows. If we remove both end-points from $[\alpha, \beta]$ the resulting space (α, β) is connected. Suppose we remove two distinct points from $(\alpha, \beta]$. If one of these is the end-point the resulting space has two components, otherwise it has three. In neither case is the resulting space connected and so we conclude that $(\alpha, \beta]$ is not homeomorphic to $[\alpha, \beta]$.

The same type of argument shows that S^1 is not homeomorphic to $\mathbb{R}$. In fact, it can be adapted to show that S^1 cannot even be embedded in $\mathbb{R}$. For suppose, to obtain a contradiction, that there exists an embedding $\phi\colon S^1 \to \mathbb{R}$. Now ϕS^1 is connected, since S^1 is connected, and so ϕS^1 is an interval. If we remove an interior point from an interval the resulting space has two components. But if we remove any point from S^1 the resulting space has just one component. Therefore S^1 cannot be homeomorphic to ϕS^1 and we have a contradiction.

For a different type of illustration we return to the theory of topological groups, and prove

Proposition 9.17

Let G be a topological group. The component C of the neutral element e is a closed normal subgroup of G.

Components are closed, which establishes the first point. To show that C is normal, let g be any element of C. Then $g^{-1}C$ is the image of C under translation by g^{-1}, and so is connected since C is connected. Moreover, $e = g^{-1} \cdot g \in g^{-1} \cdot C$ and so $g^{-1} \cdot C \subset C$. Therefore

$$C^{-1} \cdot C = \bigcup_{g \in C} g^{-1} \cdot C \subset C,$$

which shows that C is a subgroup. Now let g be any element of G. Then $g^{-1} \cdot C \cdot g$ is the image of C under conjugation by g, and so is connected since C is connected. Also $e = g^{-1} \cdot e \cdot g \in g^{-1} \cdot C \cdot g$ and so $g^{-1} \cdot C \cdot g \subset C$. Thus C is normal, as asserted.

9.3 Locally connected spaces

Definition 9.18

The topological space X is locally connected if the connected open neighbourhoods of each point form a neighbourhood base at that point.

For example, discrete spaces are locally connected since one-point sets are connected. Also trivial spaces are locally connected since the full set is connected. Since real intervals are connected the real line $\mathbb{R}$ is locally connected, and

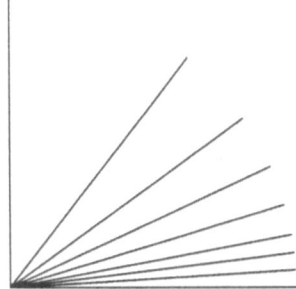

Figure 9.1. Connected subset of the real plane which is not locally connected.

similarly the real n-space $\mathbb{R}^n$ is locally connected. But not every connected space is locally connected, as illustrated by Figure 9.1.

Proposition 9.19

Let $\{X_j\}$ be a family of locally connected spaces. Suppose that X_j is connected for all but a finite number of indices j. Then the topological product $\prod X_j$ is locally connected.

For let $x = (x_j)$ be a point of the topological product and let $\prod U_j$ be a restricted product open neighbourhood of (x_j). Then $\prod U_j$ contains a connected restricted product open neighbourhood $\prod V_j$ of (x_j), where V_j is defined as follows. We take V_j to be full for those indices j for which U_j is full and connected. For the remaining indices j (necessarily finite in number) we use the assumption that X_j is locally connected and take V_j to be a connected open neighbourhood of x_j contained in U_j.

Proposition 9.20

The topological space X is locally connected if and only if the components of each open set of X are also open in X.

The condition is obviously sufficient, since if U is an open neighbourhood of a point x then the component of U containing x is a connected open neighbourhood of x. To prove the converse, let X be locally connected and let A be open in X. Let C be a component of A and let $x \in C$. If $V \subset A$ is a connected neighbourhood of x then $V \subset C$, by definition of the term component, and so C is open in X.

Proposition 9.21

Let $\pi\colon X \to X'$ be a quotient map, where X and X' are topological spaces. If X is locally connected then so is X'.

For let A' be an open set of X', so that $A = \pi^{-1}A'$ is open in X, and let C' be a component of A'. I assert that $C = \pi^{-1}C'$ is a union of components of A. For if K is the component of A containing a given point x of C then πK is connected and $\pi(x) \in \pi K \subset A'$, therefore $\pi K \subset C'$ and so $K \subset C$. Since X is locally connected and since A is open in X we have that C is open in X and so C' is open in X', as required.

It is not sufficient, in Proposition 9.21 , for π to be a continuous surjection. For example, take the integers $\mathbb{Z}$, with discrete topology. A continuous function $\rho\colon \mathbb{Z} \to \mathbb{R}$ is given by $\rho(n) = 1/n$ for $n \neq 0$ and by $\rho(0) = 0$. Although $\mathbb{Z}$ is locally connected the image $\rho\mathbb{Z}$ is not.

9.4 Pathwise-connected spaces

There is another type of connectedness, called pathwise-connectedness, which is also important and in some ways is more intuitive in nature. For a useful class of topological spaces the two types of connectedness turn out to be the same.

By a *path* in a topological space X we mean a continuous function $f\colon I \to X$, where $I = [0, 1] \subset \mathbb{R}$; we say that f *starts* at $f(0)$ and *ends* at $f(1)$. It is important to appreciate that a path is a *function*, not the image of that function.

Definition 9.22

The topological space X is pathwise-connected if for each pair of points ξ, η of X there exists a path in X which starts at ξ and ends at η.

Sierpinski spaces are pathwise-connected, more generally, so is any space with not more than three open sets. The properties of pathwise-connected spaces are similar to those of connected spaces. For example, the reader will readily prove

Proposition 9.23

Let $\phi\colon X \to Y$ be a continuous surjection, where X and Y are topological spaces. If X is pathwise-connected then so is Y.

This shows that pathwise-connectedness, like connectedness, is a topologically invariant property.

Proposition 9.24

Let $\{X_j\}$ be a family of pathwise-connected spaces. Then the topological product $\prod X_j$ is pathwise-connected.

The proof is easier than that of the corresponding result (Proposition 9.14) for ordinary connectedness. Thus let $\xi = (\xi_j)$ and $\eta = (\eta_j)$ be points of $\prod X_j$. For each index j let f_j be a path in X_j which starts at ξ_j and ends at η_j. Let f be the path in $\prod X_j$ of which the jth component is f_j. Then f starts at ξ and ends at η, as required.

Proposition 9.25

Let X be a topological space. If X is pathwise-connected then X is connected.

For suppose, to obtain a contradiction, that X is disconnected, so that there exists a continuous non-constant function $\lambda \colon X \to \mathbb{D}$, for some discrete $\mathbb{D}$. Let ξ, η be points of X such that $\lambda(\xi) \neq \lambda(\eta)$. Since X is pathwise-connected there exists a path f in X which starts at ξ and ends at η. Then $\lambda f \colon I \to \mathbb{D}$ is continuous and non-constant. Since I is connected this gives us our contradiction.

In general, the converse of Proposition 9.25 is false, as Figure 9.2 may indicate – the reader should be able to construct a proof without too much difficulty.

Pathwise-connectedness is particularly easy to study in the case of subsets of $\mathbb{R}^n$ since paths can be constructed by taking advantage of the geometry. For example, if X is a convex subspace of $\mathbb{R}^n$ then, for any points $\xi, \eta \in X$, a path f in X from ξ to η is given by

$$f(t) = (1 - t)\xi + t\eta \qquad (t \in I).$$

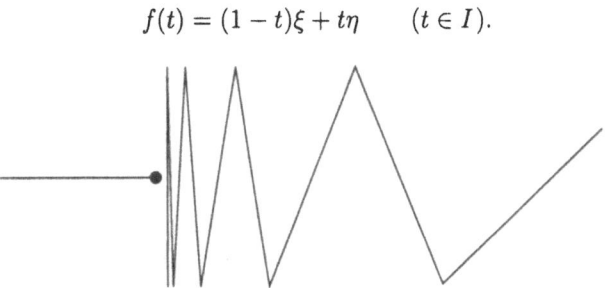

Figure 9.2. The topologists' sine curve: a connected subset of the real plane which is not pathwise-connected.

In particular, the open n-ball U^n and the closed n-ball B^n are pathwise-connected. The $(n-1)$-sphere S^{n-1} is not convex, of course, but here paths can be constructed in a similar fashion using geodesic segments rather than straight lines.

If f is a path in the topological space X the *reverse path* f^{-1} is defined by

$$f^{-1}(t) = f(1-t) \qquad (t \in I).$$

Thus f^{-1} starts where f ends and vice versa. If f, g are paths in X such that $f(1) = g(0)$ (i.e. g starts where f ends) the *juxtaposition path* h is defined by

$$h(t) = \begin{cases} f(2t) & (0 \le t \le \frac{1}{2}), \\ g(2t-1) & (\frac{1}{2} \le t \le 1), \end{cases}$$

as illustrated in Figure 9.3. To establish the continuity of h we use Corollary 3.10. Note that h starts where f starts and h ends where g ends.

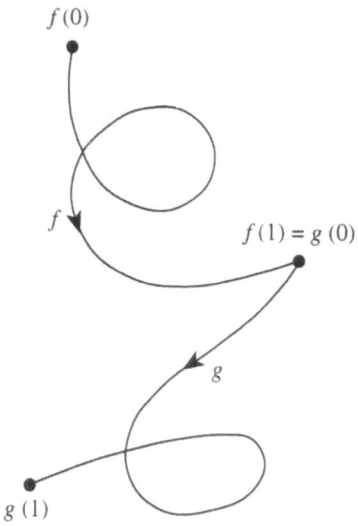

Figure 9.3. Juxtaposition of paths.

Pathwise-connectedness components (usually called path-components) can be defined as the equivalence classes determined by an equivalence relation as follows.

Definition 9.26

The points ξ, η of the topological space X are equivalent, in the sense of pathwise-connectedness, if there exists a path in X which starts at ξ and ends at η.

To establish reflexivity we use the stationary (or constant) path; to establish symmetry we use the reverse path; and to establish transitivity we use the juxtaposition of paths, as described above.

It follows from Proposition 9.25 that X has at least as many path-components as connectedness components. In general, the path-components of X are not closed (nor open, either); there is no analogue of Proposition 9.6 for pathwise-connectedness.

Definition 9.27

The topological space X is locally pathwise-connected if for each point x of X the pathwise-connected neighbourhoods of x form a neighbourhood base.

Obviously the real line $\mathbb{R}$ is locally pathwise-connected and so, more generally, is the real n-space $\mathbb{R}^n$. More generally still, any open subspace of $\mathbb{R}^n$ is locally pathwise-connected. Of course, any locally pathwise-connected space is also locally connected.

Again we see that the direct image of a locally pathwise-connected space under a continuous open surjection is also locally pathwise-connected, and hence that the property is a topological invariant.

Proposition 9.28

Let X be connected and locally pathwise-connected. Then X is pathwise-connected.

For let x be a point of X, let H be the set of points x' of X which can be joined to x by a path in X, and let $K = X - H$. I assert that both H and K are open in X. This will imply that K is empty, since X is connected, and so that X is pathwise-connected.

First, we show that H is open. For if x' is a point of H there exists an open neighbourhood U of x' such that for any point x'' of U there exists a path in U from x' to x''. But since x' is in H there is also a path in X from x to x'. Juxtaposing these two paths we obtain a path in X from x to x'', so that $x'' \in H$. Thus $U \subset H$ and so H is open.

Second, we show that K is open. For if x' is a point of K there exists an open neighbourhood U of x' such that for any point x'' of U there exists a path in U from x'' to x'. Then x'' is not in H since, otherwise, there would exist a path in X from x to x'' and hence, by juxtaposition, a path in X from x to x', contrary to the assumption that $x' \in K$. So $x'' \in K$, thus $U \subset K$ and so K is open.

Now the proof is completed as indicated in the first paragraph. The result applies, for example, to open subspaces of $\mathbb{R}^n$ and of S^n. It can readily be extended to

Proposition 9.29

Let X be a locally pathwise-connected topological space. Then the components of X and the path-components of X coincide.

For a better understanding of the relationship between the two forms of connectedness consider the diagram shown below, where X' is the quotient space of X with respect to the equivalence relation defined in Definition 9.26 while X'' is the quotient space of X with respect to the equivalence relation defined in Definition 9.15.

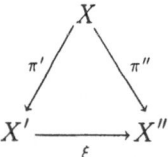

The function ξ, induced by the identity on X, is a continuous surjection in all cases. In some cases, as when X is locally pathwise-connected, ξ is also injective. Then ξ^{-1} is also continuous, from consideration of our next diagram, and so is a homeomorphism.

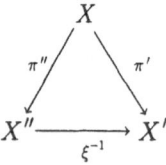

As usual, the case of a topological group G has special features. We have already seen, in Proposition 9.17, that the connectedness component of the neutral element e constitutes a normal subgroup. In fact, the set of components of G forms a group, in the algebraic sense, with the group structure inherited from G, and this group is isomorphic to the factor group of G by the component of e. A similar result holds for the path-components of G.

9.5 Uniformly connected spaces

Definition 9.30

The uniform space X is uniformly connected if for each discrete uniform space $\mathbb{D}$ every uniformly continuous function $\lambda \colon X \to \mathbb{D}$ is constant.

When X has the discrete uniformity we can take $X = \mathbb{D}$ and λ the identity; then X is not uniformly connected if X has more than one point. When X has the trivial uniformity then $(\lambda \times \lambda)^{-1} \Delta \mathbb{D} = X \times X$, since $\Delta \mathbb{D}$ is an entourage of $\mathbb{D}$; therefore, λ is constant and X is uniformly connected. This can also be seen from

Proposition 9.31

Suppose that the uniform space X is connected, in the uniform topology. then X is uniformly connected.

For let $\lambda \colon X \to \mathbb{D}$ be uniformly continuous, where $\mathbb{D}$ is discrete. Then λ is continuous and so constant, since X is connected.

As we shall see later the converse of Proposition 9.31 is false. For example, the rational line $\mathbb{Q}$ is uniformly connected, in the euclidean uniformity, although not connected, in the euclidean topology.

The properties of uniform connectedness are similar to those of topological connectedness. For example, we have

Proposition 9.32

Let $\phi \colon X \to Y$ be a uniformly continuous surjection, where X and Y are uniform spaces. If X is uniformly connected then so is Y.

Proposition 9.33

Let $\{X_j\}$ be a family of uniformly connected spaces. Then the uniform product $\prod X_j$ is uniformly connected.

Proposition 9.34

Let $\{X_j\}$ be a covering of the uniform space X by uniformly connected subspaces. Suppose that some member of the covering meets all the others. Then X is uniformly connected.

For let $\lambda\colon X \to \mathbb{D}$ be uniformly continuous, where $\mathbb{D}$ is discrete. Then $\lambda|X_j$ is uniformly continuous, for each index j, and so constant, since X_j is uniformly connected. Let a_j be the constant value of $\lambda|X_j$. Now each X_j meets X_k for some index k, and so $a_j = a_k$. Therefore λ itself is constant and so X is uniformly connected.

An alternative characterization of uniform connectedness is provided by

Proposition 9.35

The uniform space X is uniformly connected if and only if for each pair of points $\xi, \eta \in X$ and each entourage E of X we have $(\xi, \eta) \in E^n$ for some integer n.

For suppose that $\xi, \eta \in X$ and an entourage E exist such that $(\xi, \eta) \notin E^n$ for every n. Taking $\mathbb{D} = \{-1, +1\}$, define $\lambda(x) = -1$ if $(\xi, x) \in E^n$ for some n and $\lambda(x) = +1$ otherwise. Then E is contained in the inverse image of the diagonal with respect to $\lambda \times \lambda$, and so λ is uniformly continuous. Since λ is not constant this shows that X cannot be uniformly connected. This proves Proposition 9.35 in one direction.

For the proof in the other direction, let $\mathbb{D}$ be any discrete uniform space and let $\lambda\colon X \to \mathbb{D}$ be uniformly continuous. Take E to be the inverse image of the diagonal; then $E^n = E$ for all n. So if $(\xi, \eta) \in E^n$ for some n then $(\xi, \eta) \in E$ and so λ is constant, as required.

For example, let G be a topological group. Then Proposition 9.35 shows that G is uniformly connected if and only if every neighbourhood of the neutral element generates the whole of G. More generally, if H is a subgroup of G then G/H is uniformly connected if and only if for every neighbourhood W of H the union of the subsets W^n is the whole of G.

Definition 9.36

The uniform space X has property S if for each entourage D the set X can be covered by a finite family of connected D-small subsets.

The S here stands for Sierpinski.[12] Clearly it is sufficient if the condition in Definition 9.36 is satisfied for members of a uniformity base. For example, in case X is a metric space, it is sufficient if for each positive ε, X can be covered by a finite family of connected U_ε-small subsets, or equivalently, if X can be covered by a finite family of connected subsets, each of which is contained in an open 2ε-ball. Thus bounded intervals of the real line have property S, in the euclidean uniformity, and similarly in higher dimensions.

Clearly if X has property S then X is totally bounded. We also have

Proposition 9.37

If the uniform space X has property S then X is locally connected.

For let x be a point of X and let D be an entourage of X. We show that the uniform neighbourhood $D[x]$ contains a connected neighbourhood of x. Let E be a closed entourage of X such that $E \subset D$. Since X has property S there exists a family $\{C_1, \ldots, C_n\}$ of connected E-small sets covering X. Let C be the union of those members C_i of the family such that $x \in \mathrm{Cl}\, C_i$. Then C is connected. Moreover, C is a neighbourhood of x since x does not adhere to $X - C$. Finally E is closed and so

$$C \subset \bigcup C_i \subset \bigcup \mathrm{Cl}\, C_i \subset E[x] \subset D[x],$$

where the unions are taken over the same indices i as before. This proves Proposition 9.37.

9.6 Uniformly locally connected spaces

To understand property S better we also introduce

Definition 9.38

The uniform space X is uniformly locally connected if each entourage D contains an entourage E such that the uniform neighbourhood $E[x]$ is connected for each point x of X.

---

[12] See G.T. Whyburn, *Analytic Topology* (American Mathematical Society, Providence, RI, 1942) for the metric version of what follows, and P.J. Collins. On uniform connection properties, (*Amer. Math. Monthly*, **78** (1971), 372–374) for the uniform version.

Proposition 9.39

Let X be a compact Hausdorff space. If X is locally connected then X is uniformly locally connected.

Here, of course, we are referring to the unique uniformity on X which is compatible with the topology. To prove Proposition 9.39, let D be an entourage of X. There exists a symmetric entourage E of X such that $E \circ E \subset D$. Since X is locally connected and compact there exists, by Definition 9.18, a connected uniform neighbourhood $C_x[x] \subset E[x]$ for each point x of X. Now the neighbourhood

$$N = \bigcup_{x \in X} (C_x[x] \times C_x[x])$$

of the diagonal is an entourage, since X is compact Hausdorff. As a union of connected sets containing x each uniform neighbourhood $N[x]$ is connected, and we have

$$N \subset \bigcup_{x \in X} (E[x] \times E[x]) \subset E \circ E \subset D,$$

as required.

Proposition 9.40

Let X be a totally bounded uniform space. If X is locally uniformly connected then X has property S.

Given an entourage D of X let E be a symmetric entourage such that $E \circ E \subset D$. If X is uniformly locally connected E contains an entourage F such that the uniform neighbourhood $F[x]$ is connected for each point x of X. Since X is totally bounded there exists a finite family $A_1, \ldots, A_n$ of non-empty F-small sets which cover X. Choose $x_i \in A_i$ for $i = 1, \ldots, n$. Then $F[x_1], \ldots, F[x_n]$ is a finite family of connected sets covering X. Moreover, $F[x_i]$ is D-small for each index i, and so Proposition 9.40 is proved.

Thus the last three propositions show that a compact Hausdorff space is locally connected if and only if it has property S.

Exercises

9.1. The topological space X is the union of closed subspaces X_1 and X_2. If X and $X_1 \cap X_2$ are connected show that X_1 and X_2 are connected.

9.2. Show that the Sorgenfrey line L is not connected.

9.3. Show that, in the real plane $\mathbb{R} \times \mathbb{R}$ with the euclidean topology, the set of points with at least one irrational coordinate is connected. Is the same true if irrational is replaced by rational?

9.4. Let $X \subset \mathbb{R} \times \mathbb{R}$ be the subspace of points (x, y) such that either (i) x is irrational and $0 \le y \le 1$ or (ii) x is rational and $-1 \le y \le 0$. Prove that X is connected, with the euclidean topology.

9.5. Let X be a topological space which is the union of connected subspaces $A_1, \ldots, A_n$, such that the intersections $A_1 \cap A_2, \ldots,$ $A_{n-1} \cap A_n$ are all non-empty. Does it follows that X is connected?

9.6. Let $\phi \colon X \to Y$ be a quotient map such that $\phi^{-1}y$ is connected for each point y of Y. Let Y' be an open or closed subset of Y. Show that Y' is connected if and only if $\phi^{-1}Y'$ is connected.

9.7. Show that a subspace of a totally disconnected space is totally disconnected, and that the topological product of totally disconnected spaces is totally disconnected.

9.8. Show that for any open neighbourhood U of the neutral element e in the connected topological group G the subsets U^n $(n = 1, 2, \ldots)$ form an open covering of G. Deduce that if G is also finite the topology of G must be trivial.

9.9. Let H be a discrete normal subgroup of the connected topological group G. Show that H is a central subgroup.

9.10. Let X be a compact Hausdorff space. Let $\mathscr{B}$ be a filter base on X consisting of closed connected subsets of X. Show that the intersection of the members of $\mathscr{B}$ is a closed connected set.

9.11. Let X be locally connected. Let H be a subset of X and let C be connected and open in H. Show that $C = H \cap U$ for some connected open set U of X.

9.12. Show that the additive group $\mathbb{Q}$ of rationals forms a topological group with neighbourhood base at 0 consisting of the countable chain of subgroups $\ldots \supset U_{-1} \supset U_0 \supset U_1 \supset \ldots$, where

$$U_t = \{mp^t/n \colon p \nmid n\} \qquad (t = 0, \pm 1, \pm 2, \ldots).$$

Show that $\mathbb{Q}$ with this p-adic topology is totally disconnected.

9.13. Let X be the subset of the real line consisting of 0 together with the points n^{-1}, for $n = 1, 2, \ldots$. Show that X, with the euclidean topology, is totally disconnected.

9.14. Let $\phi \colon X \to Y$ be a uniformly open function, where X and Y are uniform spaces, with X non-empty and Y uniformly connected. Show that ϕ is surjective.

9.15. Let R be a compatible equivalence relation on the uniform space X such that each of the equivalence classes $R[x]$ is uniformly connected. Suppose that X/R is uniformly connected in the quotient uniform structure. Show that X is uniformly connected.

9.16. In the uniform space X points ξ and η are said to be related if for each entourage D there exists an integer n such that $(\xi, \eta) \in D^n$. Show that the equivalence classes thus defined are closed but not necessarily uniformly connected.

10
Countability and Related Topics

10.1 Countability

There are two important classes of topological spaces which satisfy countability conditions (the class of separable spaces, to be discussed later, may be regarded as a third). These are known as the first and second countability conditions, for historical reasons. However, in most accounts of the subject, including the present one, it is the second countability condition which comes first.

Definition 10.1

Let X be a topological space. If X admits a countable generating family then X is second countable.

For example, the real line $\mathbb{R}$, with the euclidean topology, is second countable, since the topology is generated by the countable family

$$\{(\alpha, \beta) \colon \alpha, \beta \in \mathbb{Q}\}.$$

(Of course, the same family with $\alpha, \beta \in \mathbb{R}$ is also generating but is non-countable.)

Note that a family Γ of subsets of X is countable if and only if the family Γ' of finite intersections of members Γ is countable. Without real loss of generality, therefore, we may assume that a second countable space admits a countable generating family which is complete, in the sense of Chapter 1.

Proposition 10.2

Let $\phi\colon X \to Y$ be a continuous open surjection, where X and Y are topological spaces. If X is second countable then so is Y.

For let Γ be a complete generating family for the topology of X. Then the direct image $\phi_*\Gamma$ is a generating family for the topology of Y. Moreover, $\phi_*\Gamma$ is countable if Γ is countable, and so Proposition 10.2 is obtained.

Proposition 10.3

Let $\phi\colon X' \to X$ be an embedding, where X and X' are topological spaces. If X is second countable then so is X'.

For the inverse image of a countable generating family for the topology of X is a countable generating family for the topology of X'. In particular, each subspace of a second countable space is second countable.

Proposition 10.4

Let $X_1, \ldots, X_n, \ldots$ be a countable family of second countable spaces. Then the topological product $X_1 \times \cdots \times X_n \times \cdots$ is second countable.

This follows since to generate the product topology it is sufficient to take all $\pi_n^{-1}(U(n, r))$, where $\{U(n, r)\}_{r=1,2,\ldots}$ is a generating family for the topology of X_n.

These results show that the class of second countable spaces includes the real n-space $\mathbb{R}^n$ and all its subspaces.

10.2 Lindelöf spaces

Next we study a condition which may be regarded as a weak form of compactness.

Definition 10.5

The topological space X is Lindelöf if each open covering of X admits a countable subcovering.

Proposition 10.6

Let $\phi: X \to Y$ be a continuous surjection, where X and Y are topological spaces. If X is Lindelöf then so is Y.

For let $\Gamma = \{V_j\}$ be an open covering of Y. The inverse image $\phi^*\Gamma = \{\phi^{-1}V_j\}$ is an open covering of X. If X is Lindelöf we can extract a countable subcovering of X from $\phi^*\Gamma$. The corresponding members of Γ form a countable subcovering of Y.

Proposition 10.7

Let $\phi: X' \to X$ be a closed embedding, where X and X' are topological spaces. If X is Lindelöf then so is X'.

For let Γ' be an open covering of X'. Then $\Gamma' = \phi^*\Gamma$, where Γ is a covering of $\phi X'$ by open sets of X. By adjoining to Γ the open set $X - \phi X'$ we obtain an open covering Γ^+ of X. If X is Lindelöf we can extract from Γ^+ a countable subcovering of X. Remove the additional set $X - \phi X'$ if it occurs in the subcovering. We are left with a countable subcovering of Γ of which the inverse image forms a countable subcovering of Γ'. This proves Proposition 10.7. In particular, closed subspaces of Lindelöf spaces are Lindelöf.

Examples can be given to show that finite topological products of Lindelöf spaces are not necessarily Lindelöf. Possibly this is the main reason why they are much less important than compact spaces, but they do have a number of attractive properties nevertheless.

Proposition 10.8

Let X be a second countable space. Then X is Lindelöf.

For let $\Gamma = \{W_n : n = 1, 2, \ldots\}$ be a countable complete generating family for the topology of X. Given an open covering $\{U_j\}$ of X consider the collection Σ of integers n such that W_n is contained in some member U_j of the covering. For each $n \in \Sigma$ choose an index j_n, say, such that $W_n \subset U_{j_n}$. Now each U_j is a union of members of the generating family Γ, therefore each point of U_j is contained in W_n for some n in Σ and hence in U_{j_n}. Thus the family $\{U_{j_n} : n \in \Sigma\}$ covers X and we have extracted a countable subcovering of the given covering, as required.

Proposition 10.9

Let X be a regular Lindelöf space. Then X is normal.

To see this, let E, F be closed sets of X which are disjoint. Then E, F have the Lindelöf property, by Proposition 10.7. Since X is regular we can choose for each point ξ of E an open neighbourhood U_ξ of which the closure does not intersect F, and for each point η of F an open neighbourhood V_η of which the closure does not intersect E. By the Lindelöf condition we can extract a countable subcovering $\{U_{\xi_1}, U_{\xi_2}, \dots\}$ of E from the open covering $\{U_\xi \colon \xi \in E\}$ and extract a countable subcovering $\{V_{\eta_1}, V_{\eta_2}, \dots\}$ of F from the open covering $\{V_\eta \colon \eta \in F\}$. Now construct open sets $S_1, S_2, \dots; T_1, T_2, \dots$ as follows:

$$S_1 = U_1, \qquad\qquad T_1 = V_1 - \operatorname{Cl} S_1,$$
$$S_2 = U_2 - \operatorname{Cl} T_1, \qquad T_2 = V_2 - \operatorname{Cl}(S_1 \cup S_2),$$
$$S_3 = U_3 - \operatorname{Cl}(T_1 \cup T_2), \qquad T_3 = V_3 - \operatorname{Cl}(S_1 \cup S_2 \cup S_3),$$

and so on (the suffixes ξ and η have been suppressed to ease the printing). Then $S = \bigcup S_n$ and $T = \bigcup T_n$ are disjoint open sets containing E and F, respectively. Thus X is normal, as asserted.

10.3 Countably compact spaces

In the history of general topology several different compactness conditions were studied before the one which is now standard was accepted as the most satisfactory. These conditions agree on a reasonably broad class of topological spaces although they differ in general. We consider just two of these conditions here, beginning with

Definition 10.10

The topological space X is countably compact if each countable open covering of X admits a finite subcovering.

Clearly X is compact, in the standard sense, if and only if X is both countably compact and Lindelöf; in a way, we have broken up the definition of compactness into two parts.

Proposition 10.11

Let $\phi\colon X \to Y$ be a continuous surjection, where X and Y are topological spaces. If X is countably compact then so is Y.

For let Γ be a countable open covering of Y. The inverse image $\phi^*\Gamma$ is a countable open covering of X. If X is countably compact we can extract from $\phi^*\Gamma$ a finite subcovering of X. The corresponding members of Γ form a finite subcovering of Y.

Proposition 10.12

Let $\phi\colon X' \to X$ be a closed embedding, where X and X' are topological spaces. If X is countably compact then so is X'.

For let Γ' be a countable open covering of X'. Then $\Gamma' = \phi^*\Gamma$, where Γ is a countable covering of $\phi X'$ by open sets of X. By adjoining to Γ the open set $X - \phi X'$ we obtain a countable open covering Γ^+ of X. If X is countably compact we can extract from Γ^+ a finite subcovering of X. Remove the additional set $X - \phi X'$ if it occurs in the subcovering. We are left with a finite subcovering of Γ of which the inverse image is a finite subcovering of Γ'. This proves Proposition 10.12. In particular, closed subspaces of countably compact spaces are countably compact.

Examples can be given to show that finite topological products of countably compact spaces are not necessarily countably compact.

There is an interesting characterization of countably compact Hausdorff spaces involving the notion of accumulation point (not to be confused with the notion of adherence point).

Definition 10.13

Let H be a subset of the topological space X. The point x of X is an accumulation point of H if each neighbourhood of x intersects $H - \{x\}$. The derived set of H is the set of accumulation points.

In other words, x is an accumulation point of H if and only if x is an adherence point of $H - \{x\}$. Whereas points of H itself are automatically adherence points this is not the case with accumulation points.

Proposition 10.14

Let X be a Hausdorff space. Then X is countably compact if and only if there exists an accumulation point for each infinite subset of X.

For suppose that X is countably compact. Clearly it is sufficient to show that for each sequence $\{x_n\colon n = 1, 2, \ldots\}$ of points of X the set $H = \{x_1, x_2, \ldots\}$ has an accumulation point. Without real loss of generality we may suppose that all the terms of the sequence are distinct. Write $H_n = \{x_n, x_{n+1}, \ldots\}$ for each n, and consider the countable family Γ^* of closed sets $\mathrm{Cl}\, H_n$ of X. I assert that the intersection of all these closed sets is non-empty: then any point of the intersection is an accumulation point of H. For suppose, to obtain a contradiction, that the intersection of the members of Γ^* is empty. Then the dual family Γ of complements is a countable open covering of X. Since X is countably compact we can extract a finite subcovering, and then the corresponding finite subfamily of Γ^* has empty intersection. Since no finite subfamily of $\{H_n : n = 1, 2, \ldots\}$ has empty intersection we have our contradiction, which proves Proposition 10.14 in one direction.

Conversely, suppose that X is not countably compact. Then there exists a countable open covering $\{U_n : n = 1, 2, \ldots\}$ of X from which it is impossible to extract a finite subcovering. For each n, therefore, we can choose a point x_n of X which does not belong to $U_1 \cup \cdots \cup U_n$. Then the set $\{x_n : n = 1, 2, \ldots\}$ has no accumulation point since each point x of X belongs to U_N for some N and yet U_N does not intersect $\{x_{N+1}, x_{N+2}, \ldots\}$. This completes the proof of Proposition 10.14.

10.4 Sequentially compact spaces

We turn now to another type of compactness which is defined in terms of convergence of sequences.

Definition 10.15

The topological space X is sequentially compact if each sequence of points of X admits a convergent subsequence.

Proposition 10.16

Let $\phi \colon X \to Y$ be a continuous surjection, where X and Y are topological spaces. If X is sequentially compact then so is Y.

For let $\langle y_n \rangle$ be a sequence of points of Y. Choose a point $x_n \in \phi^{-1}(y_n)$ for each n. Then $\langle x_n \rangle$ is a sequence of points of X. If X is sequentially compact there exists a subsequence of $\langle x_n \rangle$ which converges to some point x of X. The corresponding subsequence of $\langle y_n \rangle$ converges to the point $\phi(x)$ of Y.

Proposition 10.17

Let $\phi: X' \to X$ be a closed embedding, where X and X' are topological spaces. If X is sequentially compact then so is X'.

For let $\langle x'_n \rangle$ be a sequence of points of X'. Then $\langle x_n \rangle$ is a sequence of points of X, where $x_n = \phi(x'_n)$ for each n. If X is sequentially compact there exists a limit point x of a subsequence of $\langle x_n \rangle$. Now x is an adherence point of $\phi X'$, since each neighbourhood of x contains at least one term $\phi(x'_n)$ (in fact infinitely many). Since $\phi X'$ is closed in X this implies that $x \in \phi X'$. Then $\phi^{-1}(x)$ is a limit point of a subsequence of $\langle x'_n \rangle$. This proves Proposition 10.17; in particular, closed subspaces of sequentially compact spaces are sequentially compact.

Proposition 10.18

Let $X_1, \ldots, X_n, \ldots$ be a countable family of sequentially compact spaces. Then the topological product $X_1 \times \cdots \times X_n \times \cdots$ is sequentially compact.

A diagonal process can be used to prove this. Thus let $\langle \xi_r \rangle$ be a sequence in $X_1 \times \cdots \times X_n \times \cdots$, so that $\langle \pi_n(\xi_r) \rangle$ is a sequence in X_n for each n. Since X_n is sequentially compact there exists an injection $\alpha_n : \mathbb{N} \to \mathbb{N}$ such that $\langle \pi_n(\xi_{a_n(r)}) \rangle$ converges to a point x_n of X_n. Consider the injection $\beta : \mathbb{N} \to \mathbb{N}$ given by

$$\beta(n) = \alpha_n \alpha_{n-1} \cdots \alpha_1(n) \qquad (n \in \mathbb{N}).$$

Then $\langle \xi_{\beta(r)} \rangle$ converges in $X_1 \times \cdots \times X_n \times \cdots$ to the point x, where $\pi_n(x) = x_n$.

Proposition 10.19

Let X be a topological space. If X is sequentially compact then X is countably compact.

For let H be an infinite subset of X. Then there exists a sequence $\langle x_n \rangle$ of distinct points of H. If X is sequentially compact then $\langle x_n \rangle$ contains a subsequence $\langle x'_n \rangle$ which converges to some point x of X. Now $\langle x'_n \rangle$, like $\langle x_n \rangle$, consists of distinct points of H. Each neighbourhood of x contains an infinite number of terms of the sequence, hence at least one point of $H \cap (X - \{x\})$. Therefore x is an accumulation point of H and so X is countably compact, as asserted.

In general, the converse of Proposition 10.19 is false, but this is where the first countability axiom comes in.

Definition 10.20

Let X be a topological space. Then X is first countable if for each point x of X the neighbourhood filter $\mathcal{N}_x$ admits a countable base.

Clearly every second countable space is first countable, but not conversely; for example, a non-countable discrete space is first countable but not second countable.

In a first countable space (and *a fortiori* in a second countable space) the essential features of the topology can be expressed satisfactorily in terms of sequences. The reason for this is contained in

Proposition 10.21

Let X be a first countable space and let H be a subset of X. The point x of X adheres to H if and only if there exists a sequence $\langle x_n \rangle$ of points of H which converges to x.

For if x is an adherence point of H, choose a countable neighbourhood base $\{W_n : n = 1, 2, \ldots\}$ at x in X. Replacing W_n by $W_1 \cup \cdots \cup W_n$ if necessary we can assume that the nesting condition is satisfied. Since W_n intersects H for each n we can choose a point x_n of the intersection $H \cap W_n$. The result is a sequence $\langle x_n \rangle$ of points of H which converges to x by the nesting condition.

Conversely, suppose that $\langle x_n \rangle$ is a sequence of points of H which converges to the point x. Then each neighbourhood of x contains a term of the sequence and so intersects H. Therefore x is an adherence point of H, which completes the proof.

We go on to show that continuity can also be characterized in terms of convergence of sequences for first countable domains. This needs the preliminary

Lemma 10.22

Let X be a first countable space. Let $\mathcal{F}$ be a filter on X which admits a countable filter base and let x be an adherence point of $\mathcal{F}$. Then there exists an elementary filter on X refining $\mathcal{F}$ which converges to x.

For let $\{M_n : n = 1, 2, \ldots\}$ be a countable base for $\mathcal{F}$ and let $\{W_n : n = 1, 2, \ldots\}$ be a countable base for the neighbourhood filter $\mathcal{N}_x$ of x. We may assume, as in the previous proof, that both of these countable bases are nested. Since x is an adherence point of $\mathcal{F}$ each member of $\mathcal{N}_x$ intersects every member of $\mathcal{F}$. In particular, the intersection $W_n \cap M_n$ is non-empty for each n. Choose a point x_n of the intersection, for each n. Then the elementary filter associated with the sequence $\langle x_n \rangle$ satisfies the requirements.

With the aid of Lemma 10.22 we can prove the converse of Corollary 2.7 and so characterize continuity, for first countable domains, as follows.

Proposition 10.23

Let $\phi \colon X \to Y$ be a function, where X and Y are topological spaces with X first countable. Then ϕ is continuous at the point x of X if for each sequence $\langle x_n \rangle$ in X which converges to x, the corresponding sequence $\langle \phi(x_n) \rangle$ in Y converges to $\phi(x)$.

For by Lemma 10.22 the neighbourhood filter $\mathcal{N}_x$ can be refined by an elementary filter $\mathcal{F}_x$ which converges to x. Then $\phi_* \mathcal{N}_x$ is refined by $\phi_* \mathcal{F}_x$. Since $\phi_* \mathcal{F}_x$ converges to $\phi(x)$, by hypothesis, so does $\phi_* \mathcal{N}_x$, hence ϕ is continuous at x, by Proposition 2.5.

Proposition 10.24

Let X be a countably compact and first countable space. Then X is sequentially compact.

For suppose, to obtain a contradiction, that $\langle x_n \rangle$ is a sequence of points of X which contains no convergent subsequence. Then the set $H = \{x_n\}$ is certainly infinite and so has an accumulation point x, say, since X is countably compact. Choose a countable neighbourhood base $\{W_n \colon n = 1, 2, \ldots\}$ for x in X. Choose a point $x_{n_1} \in H \cap W_1$, other than x, a point $x_{n_2} \in H \cap W_2$, other than x, and so on. We obtain a subsequence of $\langle x_n \rangle$ which obviously converges to x.

By combining the results of the last few pages we reach the conclusion that for second countable Hausdorff spaces the conditions compact, countably compact and sequentially compact are coincident.

10.5 Separable spaces

Definition 10.25

Let X be a topological space. If X admits a countable dense subset then X is separable.

For example, the rationals $\mathbb{Q}$ form a countable dense subset of the real line $\mathbb{R}$, with the euclidean topology. More generally, any second countable space is separable, since by taking one point from each non-empty member of a countable generating family we obtain a countable dense subset.

Proposition 10.26

Let $\phi\colon X \to Y$ be a continuous surjection, where X and Y are topological spaces. If X is separable then so is Y.

For if H is a countable dense subset of X then ϕH is a countable dense subset of Y, using Proposition 2.3.

Proposition 10.27

Let $\phi\colon X' \to X$ be an open embedding, where X and X' are topological spaces. If X is separable then so is X'.

For if H is a countable dense subset of X then $\phi^{-1}H$ is a countable dense subset of X'. In particular, open subspaces of separable spaces are separable.

Proposition 10.28.

Let $X_1, \ldots, X_n, \ldots$ be a countable family of separable spaces. Then the topological product $X_1 \times \cdots \times X_n \times \cdots$ is separable.

If for each n, $\{x_1^n, x_2^n, \ldots, x_{r_n}^n, \ldots\}$ is dense in X_n, let S be the set of all sequences

$$(x_{r_1}^1, x_{r_2}^2, \ldots, x_{r_n}^n, \ldots)$$

such that $r_n = 1$ (say) for all n sufficiently large. This is countable, and every restricted product open set of $\prod X_n$ contains a point of S.

Proposition 10.29

Let X be a countably compact uniform space. Then X is totally bounded.

For suppose X is not totally bounded. Then for some entourage D there exists no finite subset S of X such that $D[S] = X$. Choose any point x_1 of X. Since

$D[x_1]$ is a proper subset of X we can choose a point x_2 of X such that $x_2 \notin D[x_1]$. Since $D[x_1, x_2]$ is a proper subset of X we can choose a point x_3 of X such that $x_3 \notin D[x_1, x_2]$. We now proceed by induction.

Suppose we have chosen a set of points $\{x_1, \ldots, x_n\}$ of X such that $(x_i, x_j) \notin D$ whenever $i \neq j$. Since $D[x_1, \ldots, x_n]$ is a proper subset of X there exists a point x_{n+1} of X such that $x_{n+1} \notin D[x_1, \ldots, x_n]$, i.e. such that $(x_{n+1}, x_i) \notin D$ for $i = 1, \ldots, n$. So by induction we have obtained a sequence $\langle x_n \rangle$ of distinct points of X such that $(x_i, x_j) \notin D$ whenever $i \neq j$.

Now the subset $H = \{x_1, x_2, \ldots\}$ of X, being countable, has an accumulation point x, say. There exists an entourage E of X such that $E \circ E \subset D$, and then $E[x]$ intersects H in at least two points x_i, x_j, say, where $i \neq j$. But then $(x_i, x_j) \in E \circ E \subset D$, contrary to the construction. From this contradiction we conclude that X is totally bounded.

Every metric space X satisfies the first condition of countability. To be sure, we have described the metric topology using as a neighbourhood base at each point x the family of open ε-balls $U_\varepsilon(x)$ for all positive ε. However, another neighbourhood base, in the same topology, consists of the family of open ε-balls $U_\varepsilon(x)$ for all positive *rational* ε, and this neighbourhood base is countable.

It is easy to give examples of metric spaces which fail to satisfy the second condition of countability, the existence of a countable generating family for the topology. For example, an uncountable set with the discrete metric cannot be second countable. We prove

Proposition 10.30

Let X be a separable metric space. Then X is second countable.

For let $H = \{x_n : n = 1, 2, \ldots\}$ be a countable dense subset of X. I claim that the countable family $\{U_{1/k}(x_n)\}$, where n, k run through the natural numbers $\mathbb{N}$, generates the metric topology. It is sufficient to show that for each point x of X and positive ε we can choose n and k so that $U_{1/k}(x_n)$ is contained in $U_\varepsilon(x)$. Choose $k > 2/\varepsilon$. Since H is dense in X there exists an integer n such that $x_n \in U_{1/k}(x)$, and then $x \in U_{1/k}(x_n)$ by symmetry of the metric ρ. Now if $\xi \in U_{1/k}(x_n)$ then

$$\rho(x, \xi) \leq \rho(x, x_n) + \rho(x_n, \xi) < 2/k < \varepsilon,$$

so that $\xi \in U_\varepsilon(x)$. Thus $U_{1/k}(x_n) \subset U_\varepsilon(x)$, as required.

Proposition 10.31

Let X be a countably compact metric space. Then X is separable.

For by Proposition 10.29 X is totally bounded, and so there exists a finite $1/n$-net for each natural number n. Let us denote this net by

$$\{x_1^n, x_2^n, \ldots, x_{k(n)}^n\}.$$

Now consider the subset

$$H = \bigcup_{n \in \mathbb{N}} \bigcup_{i=1}^{k(n)} \{x_i^n\},$$

which is certainly countable. I assert that H is dense in X. For let x be a point of X, and let $\varepsilon > 0$. Choose an integer $n > 1/\varepsilon$ and consider the $1/n$-net, as above. For some $i \geq k(n)$ we have $x \in U_{1/n}(x_i^n)$ and so $x_i^n \in U_{1/n}(x) \subset U_\varepsilon(x)$, as required.

Exercises

10.1. Let X be a first countable space. Show that X is Hausdorff if and only if each convergent sequence in X has a unique limit.

10.2. Let X be a separable space. Show that every family of pairwise disjoint open sets of X is countable.

10.3. Let X be a Lindelöf space. Show that there exists an accumulation point for each uncountable subset of X.

10.4. Show that the Sorgenfrey plane is separable, but the antidiagonal subspace is not.

10.5. Show that if X is compact and Y is a Lindelöf space then $X \times Y$ is a Lindelöf space.

10.6. Show that the Sorgenfrey line has the Lindelöf property, but the Sorgenfrey plane does not.

10.7. Let $\phi \colon X \to Y$ be a function, where X is first countable and Y is countably compact. If the graph function $\Gamma_\phi \colon X \to X \times Y$ is closed show that ϕ is continuous.

10.8. Let X be a topological space. Show that X is countably compact if and only if each closed subset of X which is discrete in the relative topology is a finite set.

10.9. Show that the uniform space X is totally bounded if and only if each of its countable subsets is totally bounded.

11
Functional Separation Conditions

11.1 General remarks

Let H, K be a disjoint pair of subsets of the topological space X. Chapter 6 is concerned with the question of whether H and K can be separated: Do there exist open neighbourhoods U, V of H, K, respectively, which are also disjoint? The present chapter is concerned with the question of whether H and K can be functionally separated: Does there exist a continuous real-valued function $\alpha \colon X \to I$ such that $\alpha = 0$ throughout H and $\alpha = 1$ throughout K? The latter condition implies the former since we can take $U = \alpha^{-1}[0, \frac{1}{2}), V = \alpha^{-1}(\frac{1}{2}, 1]$.

For example, consider the Hausdorff condition, in which H and K are one-point subsets. We may describe X as *functionally Hausdorff* if for each pair ξ, η of distinct points of X there exists a continuous real-valued function $\alpha \colon X \to I$ such that $\alpha(\xi) = 0$ and $\alpha(\eta) = 1$. We will not dwell on this condition, since it is not particularly relevant to what we are going to do, but the reader may be interested to show that subspaces of functionally Hausdorff spaces are functionally Hausdorff, also that topological products of functionally Hausdorff spaces are functionally Hausdorff.

11.2 Completely regular spaces

Next consider the regularity condition. The standard term for functionally regular is completely regular:

Definition 11.1

The topological space X is completely regular if for each closed set H of X and each point x of $X - H$ there exists a continuous real-valued function $\alpha\colon X \to I$ such that $\alpha = 0$ throughout H and $\alpha = 1$ at x.

Proposition 11.2

Let $\phi\colon X' \to X$ be an embedding, where X and X' are topological spaces. If X is completely regular then so is X'.

In particular, subspaces of completely regular spaces are completely regular. To prove Proposition 11.2, let H' be a closed set of X' and let $x' \in X' - H'$. Write $x = \phi(x')$ and $H = \mathrm{Cl}\,(\phi|H')$. Then $H' = \phi^{-1}H$, hence $x \notin H$. When X is completely regular there exists a continuous function $\alpha\colon X \to I$ such that $\alpha = 0$ throughout H and $\alpha = 1$ at x. Then $\alpha' = \alpha \circ \phi\colon X' \to I$ is a continuous function such that $\alpha' = 0$ throughout H' and $\alpha' = 1$ at x', as required.

Proposition 11.3

Let $\{X_j\}$ be a family of completely regular spaces. Then the topological product $X = \prod X_j$ is completely regular.

Let $x = (x_j)$ be a point of X and let H be a closed set of X which does not contain x. Let $\prod U_j$, where U_j is open in X_j, be a restricted product open neighbourhood of x which does not intersect H. Thus U_j is full for all but a finite number of values of j, say $j(1), \ldots, j(n)$. Given $i = 1, \ldots, n$ there exists a continuous function $\alpha_i\colon X_{j(i)} \to I$ such that $\alpha_i = 0$ throughout the closed set $X - U_{j(i)}$ while $\alpha_i = 1$ at $x_{j(i)}$. The continuous function $\beta_i = \alpha_i \pi_{j(i)}\colon X \to I$ is zero away from $\pi_{j(i)}^{-1} U_{j(i)}$. Using multiplication of real numbers, therefore, we can define a continuous function $\alpha\colon X \to I$ by

$$\alpha(x) = \beta_1(x) \ldots \beta_n(x) \qquad (x \in X),$$

and $\alpha = 0$ throughout $H \subset X - \prod U_j$ while $\alpha = 1$ at x, as required.

Proposition 11.4

The topological space X is completely regular if and only if the topology of X is generated by the family of cozero sets $\alpha^{-1}(0,1]$ of continuous functions $\alpha\colon X \to I$.

For suppose that X is completely regular. Let x be a point of X and let U be an open neighbourhood of x. Then there exists a continuous function $\alpha \colon X \to I$ such that $\alpha = 0$ throughout $X - U$ and $\alpha = 1$ at x. The cozero set $\alpha^{-1}(0,1]$ contains x and is contained in U. Since the cozero sets are open and cover X they generate the topology.

Conversely, suppose that the family of cozero sets generates the topology. Let H be a closed set of X and let x be a point of $X - H$. Since $X - H$ is an open neighbourhood of x there exists a continuous function $\beta \colon X \to I$ such that $\beta = 0$ throughout H and $\beta \neq 0$ at x. By postcomposing β with an appropriate self-homeomorphism of I we obtain a continuous function $\alpha \colon X \to I$ such that $\alpha = 0$ throughout H and $\alpha = 1$ at x. Thus X is completely regular, as asserted.

11.3 Uniformizability

We now come to the first major result of this chapter, the proof of which involves a different idea from any we have encountered so far.

Proposition 11.5

The topological space X is uniformizable if and only if X is completely regular.

By uniformizable we mean, of course, that there exists a uniformity such that the associated uniform topology is the given topology. We have already shown that uniform spaces are regular, and examples exist of regular spaces which are not uniformizable; the functional separation axiom is essential for this result.

In fact, any topological space X can be given uniform structure as follows: give X the coarsest uniformity such that $\alpha \colon X \to I$ is uniformly continuous, for each continuous function α, with the euclidean uniformity on I. Thus a base for the entourages consists of finite intersections of subsets of $X \times X$ of the form

$$D_{\alpha,\varepsilon} = \{(\xi,\eta) \in X \times X \colon |\alpha(\xi) - \alpha(\eta)| < \varepsilon\},$$

where $\alpha \colon X \to I$ is continuous and ε is positive. For each point x of X the subset $D_{\alpha,\varepsilon}[x]$ is open in the given topology, since

$$D_{\alpha,\varepsilon}[x] = \{\xi \colon |\alpha(\xi) - \alpha(x)| < \varepsilon\} = \alpha^{-1}(\alpha(x) - \varepsilon, \alpha(x) + \varepsilon).$$

So the associated uniform topology is no finer than the given topology.

To prove Proposition 11.5, first suppose that X is completely regular. Then the uniform topology just defined is no coarser than the given topology. For let H be a closed set of the given topology and let x be a point of $X - H$. Since X is completely regular there exists a continuous function $\alpha: X \to I$ such that $\alpha = 0$ throughout H and $\alpha = 1$ at x. Consider the entourage

$$D = D_{\alpha,1/2} = \{(\xi, \eta) \in X \times X: |\alpha(\xi) - \alpha(\eta)| < \tfrac{1}{2}\}.$$

If $\xi \in D[x]$ we have $(x, \xi) \in D$ so that $|\alpha(x) - \alpha(\xi)| < \tfrac{1}{2}$ and hence $|\alpha(\xi)| > \tfrac{1}{2}$. It follows that the uniform neighbourhood $D[x]$ does not intersect H and so H is closed in the uniform topology. We conclude, therefore, that the uniform topology coincides with the given topology. This proves Proposition 11.5 in one direction.

In the other direction we suppose that X is a uniform space, and use a variant of an ingenious construction for non-constant continuous functions on X due to Urysohn, as follows. Let H be a closed set of X, in the uniform topology, and let x be a point of $X - H$. Then $D[x] \subset X - H$, for some entourage D of X. Write $D = D_0$ and then, proceeding inductively, let $D_0, \ldots, D_i, \ldots$ be entourages of X such that $D_{i+1} \circ D_{i+1} \subset D_i$ $(i = 0, 1, \ldots)$.

Each positive real number $t \in I$ can be expressed in binary notation, thus

$$t = a_0 + a_1/2 + \cdots + a_i/2^i + \cdots,$$

where each digit a_i is either 0 or 1. Consider the subset I' of I consisting of those $t \in I$ for which the binary expansion terminates, i.e. the dyadic rationals of I. We associate an entourage E_t of X with each such t as follows: if $i_1, i_2, \ldots, i_n$ are the indices i for which $a_i = 1$, in the expansion of t, with $i_1 < i_2 < \cdots < i_n$, then

$$E_t = D_{i_n} \circ D_{i_{n-1}} \circ \cdots \circ D_{i_1}.$$

In addition, we set $E_0 = \Delta X$, the diagonal.

I assert that if $s, t \in I'$ with $s \leq t$ then $E_s \subset E_t$. This is easy to see in case s, t are of the form $k/2^n$, $(k+1)/2^n$, respectively, since then $D_n \circ E_s \subset E_t$. Hence, by induction, we obtain the assertion in the special case when s, t are of the form $k/2^n$, $l/2^n$, with $l > k$. The general case now follows by taking n to be the last index i in the binary expansions of s and t for which the digit a_i is non-zero, and putting $k = 2^n s$, $l = 2^n t$.

We now define our function $\alpha: X \to I$ by

$$\alpha(\xi) = \inf\{t \in I': \xi \notin E_t[x]\}$$

when $\xi \neq x$, and by $\alpha(x) = 1$. If $\xi \in H$ then $\xi \notin E_0[x]$, since $E_0 = \Delta X$, and so $\alpha(\xi) = 0$. All that remains to be shown is that α is continuous.

In fact, α is uniformly continuous. For let n be any positive integer; I assert that if $(\xi, \eta) \in D_n$ then

$$|\alpha(\xi) - \alpha(\eta)| < 1/2^n.$$

For observe that if $s = k/2^n$ and $t = (k+1)/2^n$ then $\alpha(\xi) < s$ implies $\alpha(\eta) \le t$, since

$$\xi \in E_s[x] \Rightarrow \eta \in (D_n \circ E_s)[x] \subset E_t[x],$$

while $\alpha(\eta) < s$ implies $\alpha(\xi) \le t$ similarly. Thus the interval between $\alpha(\xi)$ and $\alpha(\eta)$ cannot contain an interval $[s, t]$ of the above form, and so the assertion follows. This establishes continuity and completes the proof of Proposition 11.5.

It is noteworthy that a completely regular T_1 space can be embedded as a subspace of a compact Hausdorff space, specifically, as a subspace of the topological Jth power I^J of the closed unit interval I for an appropriate indexing set J. In fact, for any topological space X we can take J to be the set of continuous functions $\phi_j \colon X \to I$ and consider the continuous function

$$\Phi \colon X \to I^J$$

defined by $\pi_j \Phi = \phi_j$ $(j \in J)$. Suppose that X is T_1 and completely regular; then Φ is injective, as well as continuous, and it is not difficult to see that Φ is an embedding, as follows.

We have to show that ΦU is open in ΦX whenever U is open in X. So let y_0 be any point of ΦU and let x_0 be a point of X such that $\Phi(x_0) = y_0$. Choose an index j such that $\phi_j = 0$ on $X - U$ and $\phi_j > 0$ at x_0. Then $V = \pi_j^{-1}(0, 1]$ is open in I^J and so $W = V \cap \Phi X$ is open in ΦX. Now $y_0 \in W$ since $\pi_j(y_0) = \pi_j \Phi(x_0) = \phi_j(x_0) > 0$. Also $W \subset \Phi U$. For if $y \in W$ then $y = \Phi(x)$ for some $x \in X$ and $\pi_j(y) \in (0, 1]$. But $\pi_j(y) = \pi_j \Phi(x) = \phi_j(x)$, and $\phi_j = 0$ away from U. So $x \in U$ and $\Phi(x) \in \Phi U$, as required.

11.4 The Urysohn theorem

Functional normality is defined by analogy with Definition 11.1.

Proposition 11.6

Let X be a normal space. Then X is functionally normal.

Thus X is normal if and only if X is functionally normal, consequently the latter term can be dropped once this result has been established. It is usual to refer to Proposition 11.6 as Urysohn's lemma: the proof has features in common with the proof of Proposition 11.5 just given.

Let H, K be disjoint closed sets of the normal space X; we have to show that there exists a continuous function $\alpha: X \to I$ such that $\alpha = 0$ throughout H and $\alpha = 1$ throughout K, and again we use a dyadic procedure. Since X is normal there exists an open set $U_{1/2}$ such that

$$H \subset U_{1/2}, \qquad \mathrm{Cl}\, U_{1/2} \subset X - K.$$

Now H, $X - U_{1/2}$ are disjoint closed sets and $\mathrm{Cl}\, U_{1/2}$, K are disjoint closed sets. Again since X is normal there exist open sets $U_{1/4}$ and $U_{3/4}$ such that

$$H \subset U_{1/4}, \qquad \mathrm{Cl}\, U_{1/4} \subset U_{1/2}, \qquad \mathrm{Cl}\, U_{1/2} \subset U_{3/4}, \qquad \mathrm{Cl}\, U_{3/4} \subset X - K,$$

as illustrated in Figure 11.1. Clearly the process can be repeated again and again.

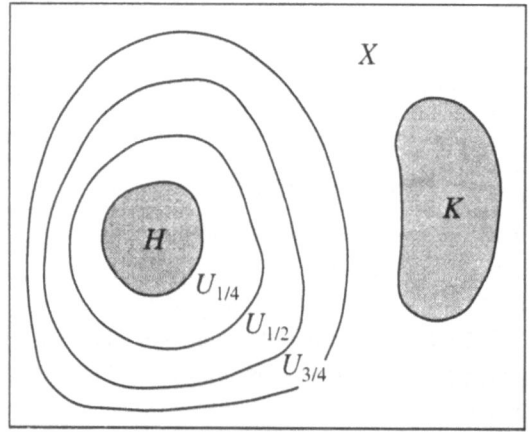

Figure 11.1.

Suppose, therefore, that for some n there exist open sets $U_{k/2^n}$ ($k = 1, \ldots,$ $2^n - 1$) satisfying the conditions

$$H \subset U_{1/2^n}, \ldots, \mathrm{Cl}\, U_{(k-1)/2^n} \subset U_{k/2^n}, \ldots, \mathrm{Cl}\, U_{(2^n-1)/2^n} \subset X - K.$$

Since X is normal there exist open sets $U_{k/2^{n+1}}$ ($k = 1, 3, 5, \ldots, 2^{n+1} - 1$) satisfying the corresponding conditions with $n + 1$ in place of n. By induction, therefore, we conclude that for each dyadic rational $t \in I'$ there exists an open set U_t such that

$$H \subset U_t, \qquad \mathrm{Cl}\, U_t \subset X - K.$$

Furthermore, the open sets can be chosen so that $\mathrm{Cl}\, U_s \subset U_t$, whenever $s < t$, for $s, t \in I'$.

Now we define $\alpha\colon X \to I$ by

$$\alpha(\xi) = \inf\{t \in I' : \xi \in U_t\}$$

if ξ belongs to some U_t, otherwise $\alpha(\xi) = 1$. Clearly $\alpha = 0$ throughout H and $\alpha = 1$ throughout K. All that remains to be shown is that α is continuous. In fact, continuity at points ξ where $0 < \alpha(\xi) < 1$ follows from the observation that $s \le \alpha(\xi) \le t$ is implied by $\xi \in U_t - \mathrm{Cl}\, U_s$; continuity at points ξ where $\alpha(\xi) = 0$ follows from the observation that $\alpha(\xi) \le t$ for $\xi \in U_t$; and continuity at points ξ where $\alpha(\xi) = 1$ follows from the observation that $\alpha(\xi) \ge t$ for $\xi \notin U_t$. This establishes Proposition 11.6.

11.5 The Tietze theorem

Our final result, which depends very much on the Urysohn lemma, can be stated in two forms.

Proposition 11.7

Let A be a closed subspace of the normal space X. Then

(i)   *each continuous function $\phi\colon A \to [-1, 1]$ admits a continuous extension $\psi\colon X \to [-1, 1]$, and*

(ii) *each continuous function $\phi\colon A \to \mathbb{R}$ admits a continuous extension $\psi\colon X \to R$.*

It is (ii) here which is generally known as the Tietze theorem. In fact, (ii) is a consequence of (i). To see this, observe first of all that $\mathbb{R}$ can be replaced by the open interval $(-1, 1)$ in (ii). Suppose, therefore, that ϕ has values in $(-1, 1)$. Apply (i) to the inclusion $\phi'\colon A \to [-1, 1]$ of ϕ and obtain a continuous extension $\psi'\colon X \to [-1, 1]$ of ϕ'. Write $B = \psi'^{-1}\{-1, 1\}$. Then A, B are disjoint closed sets and so, since X is normal, there exists a continuous function $\alpha\colon X \to I$ such that $\alpha = 0$ throughout B and $\alpha = 1$ throughout A. Using real multiplication define $\psi\colon X \to (-1, 1)$ by $\psi(\xi) = \alpha(\xi) \cdot \psi'(\xi)$. Then ψ is a continuous extension of ϕ, as required.

Thus (i) implies (ii). Before starting the proof of (i) we observe that the converse of (i) is also true. For let H, K be disjoint closed sets of X. Then $H \cup K$ is a closed set of X and the function $\phi\colon H \cup K \to [-1, 1]$ is continuous, where $\phi = -1$ throughout H and $\phi = 1$ throughout K. So if ψ is a continuous extension of ϕ to X then $\alpha = \frac{1}{2}(\phi + 1)$ functionally separates H, K.

To prove Proposition 11.7(i), and hence Proposition 11.7(ii), let $\phi\colon A \to [-1,1]$ be continuous. The disjoint sets

$$H_1 = \phi^{-1}[-1,-\tfrac{1}{3}], \qquad K_1 = \phi^{-1}[\tfrac{1}{3},1]$$

are closed in A and so closed in X. By Proposition 11.6 (applied with $[-\tfrac{1}{3},\tfrac{1}{3}]$ as codomain instead of $[0,1]$) there exists a continuous function $\alpha_1\colon X \to [-\tfrac{1}{3},\tfrac{1}{3}]$ such that $\alpha_1 = -\tfrac{1}{3}$ throughout H_1 and $\alpha_1 = \tfrac{1}{3}$ throughout K_1. Clearly we have

$$|\phi(\xi) - \alpha_1(\xi)| < \tfrac{2}{3}$$

for each point ξ of A, in other words, $\phi - \alpha_1$ sends A into the interval $[-\tfrac{2}{3},\tfrac{2}{3}]$.

Now we repeat the process with $\phi - \alpha_1$ in place of ϕ. Specifically, we divide $[-\tfrac{2}{3},\tfrac{2}{3}]$ into thirds, at the points $-\tfrac{2}{9}$ and $\tfrac{2}{9}$, and apply Proposition 11.6 to the disjoint closed sets

$$H_2 = \{\xi \in A\colon \phi(\xi) - \alpha_1(\xi) \le -\tfrac{2}{9}\},$$

$$K_2 = \{\xi \in A\colon \phi(\xi) - \alpha_1(\xi) \ge \tfrac{2}{9}\}.$$

We obtain a continuous function $\alpha_2\colon X \to [-\tfrac{2}{9},\tfrac{2}{9}]$ such that $\alpha_2 = -\tfrac{2}{9}$ throughout H_2 and $\alpha_2 = \tfrac{2}{9}$ throughout K_2.

Repeating the process again and again we obtain a sequence $\alpha_1, \alpha_2, \ldots$ of continuous real-valued functions on X such that $|\alpha_k(\xi)| \le \tfrac{1}{2}(\tfrac{2}{3})^k$ when $\xi \in X$ and

$$\left| \phi(\xi) - \sum_{k=1}^{n} \alpha_k(\xi) \right| \le (\tfrac{2}{3})^n \qquad (n = 1, 2, \ldots)$$

when $\xi \in A$. We may therefore define

$$\psi(\xi) = \sum_{k=1}^{\infty} \alpha_k(\xi) \qquad (\xi \in X).$$

Clearly ψ is an extension of ϕ, and so it only remains to establish that ψ is continuous, specifically continuous at each point ξ of X.

Given $\varepsilon > 0$, let N be an integer such that

$$\sum_{k=n}^{\infty} (\tfrac{2}{3})^k < \varepsilon/2.$$

Since each of the functions α_k is continuous there exists an open neighbourhood U_k of ξ such that

$$|\alpha_k(x) - \alpha_k(\xi)| < \varepsilon/2N \qquad (x \in U_k).$$

Then if $x \in U_1 \cap \cdots \cap U_N$ we have

$$|\psi(x) - \psi(\xi)| \leq \sum_{k=1}^{N} |\alpha_k(x) - \alpha_k(\xi)| + \sum_{k>N}^{\infty} (\tfrac{2}{3})^k < N \cdot \varepsilon/2N + \varepsilon/2 = \varepsilon,$$

and so ψ is continuous at ξ.

One further remark: Simply by taking components the Tietze theorem can be generalized from the case when $\mathbb{R}$ is the codomain to the case when $\mathbb{R}^n$ is the codomain, for any value of n.

Exercises

11.1. Let X be a completely regular space. Let E, F be disjoint subsets of X, with E closed and F compact. Show that there exists a continuous function $\alpha \colon X \to I$ such that $\alpha = 0$ throughout E and $\alpha = 1$ throughout F.

11.2. Show that a completely regular space X is compact if and only if every family Γ of continuous real-valued functions on X, such that every finite subfamily has a common zero, has a common zero.

11.3. Show that if X is normal and regular then X is completely regular.

11.4. Let $\{U_1, \ldots, U_n\}$ be a finite open covering of the normal space X. Show that there exists a family $\{\phi_1, \ldots \phi_n\}$ of continuous functions $\phi_i \colon X \to I$, such that

$$\phi_1(x) + \cdots + \phi_n(x) = 1 \qquad \text{for each } x \in X,$$

$$\mathrm{Cl}\ \phi_i^{-1}(0, 1] \subset U_i \qquad \text{for each } i.$$

[*Hint*: First show that the given covering can be shrunk to an open covering $\{V_1, \ldots, V_n\}$ of X such that $\mathrm{Cl}\ V_i \subset U_i$ for each i.]

11.5. Let $\phi \colon A \to S^n$ be continuous where A is a closed subspace of the normal space X. Show that there exists a neighbourhood U of A in X and a continuous function $\psi \colon U \to S^n$ such that $\psi | A = \phi$.

11.6. Consider the subset X of $\mathbb{R}^2$ formed by the union of the point-pair $\{(0,0), (1,0)\}$ and the open unit square $(0,1) \times (0,1) = D$. Give X the topology in which neighbourhoods of points of D are the euclidean neighbourhoods, in which open neighbourhoods of $(0, 0)$ are the sets

$$U_m = \{(0,0)\} \cup \{(x,y) \colon 0 < x < \tfrac{1}{2}, 0 < y < 1/m\} \qquad (m \in \mathbb{N})$$

and in which open neighbourhoods of $(1, 0)$ are the sets

$$V_n = \{(1,0)\} \cup \{(x,y) : \tfrac{1}{2} < x < 1, 0 < y < 1/n\} \qquad (n \in \mathbb{N})$$

Show that X is a Hausdorff space but $(0,0)$ and $(1,0)$ do not have disjoint closed neighbourhoods, so that X cannot be functionally Hausdorff.

12

Completeness and Completion

12.1 Complete uniform spaces

We have seen, in Proposition 8.26, that for filters on a uniform space convergence implies the Cauchy condition, while the converse implication is generally false. This suggests

Definition 12.1

The uniform space X is complete if each Cauchy filter on X is convergent.

For example, the discrete uniformity is always complete and so is the trivial uniformity, for rather different reasons.

Proposition 12.2

If the uniform space X is complete then each closed subspace of X is complete.

For let A be a closed subspace of X with inclusion $\sigma \colon A \to X$. If $\mathscr{F}$ is a Cauchy filter on A then the extension $\sigma_* \mathscr{F}$ of $\mathscr{F}$ to X also satisfies the Cauchy condition. If X is complete then $\sigma_* \mathscr{F}$ converges to some point x of X. This implies that x is an adherence point of A, since $\mathscr{F}$ consists of subsets of A, and so $x \in A$ since A is closed. Therefore $\mathscr{F}$ converges to x and so A is complete.

In the other direction, we have

Proposition 12.3

Let X be a separated uniform space. Then each complete subspace of X is closed.

For suppose, to obtain a contradiction, that A is a complete subspace of X but not closed. Let x be an adherence point of A such that $x \notin A$. The neighbourhood filter $\mathcal{N}_x$ of x in X is a Cauchy filter on X. Hence, by Proposition 8.30, the trace of $\mathcal{N}_x$ on A is a Cauchy filter on A. Since A is complete the trace converges to some point $a \in A$. But then a would be a limit point of $\mathcal{N}_x$ and so $a = x$, since X is separated. This gives us our contradiction.

Proposition 12.4

Let A be a dense subset of the uniform space X. Then X is complete if and only if for each Cauchy filter on A the extension to X is convergent.

If X is complete the extension of a Cauchy filter on A is a Cauchy filter on X and so is convergent. To prove the converse, let $\mathcal{F}$ be a Cauchy filter on X. Consider the inclusion $\sigma \colon A \to X$. There is no reason why the trace of $\mathcal{F}$ on A should be a filter, since there may be members of $\mathcal{F}$ which do not intersect A. However, consider the filter $\mathcal{G}$ on X generated by subsets of the form $D[M]$, where D is an entourage and M is a member of $\mathcal{F}$. Since $\mathcal{F}$ is finer than $\mathcal{G}$ it will be sufficient if we can show that $\mathcal{G}$ converges. Now each set $D[M]$ is a neighbourhood of M and so intersects A; therefore, $\sigma^* \mathcal{G}$ is defined as a filter on A. Since $\mathcal{G}$ is a Cauchy filter so is $\sigma^* \mathcal{G}$, by Proposition 8.30, and hence so is $\sigma_* \sigma^* \mathcal{G}$, by Proposition 8.29. Therefore $\sigma_* \sigma^* \mathcal{G}$ is convergent, by the condition, and so has an adherence point. But $\sigma_* \sigma^* \mathcal{G}$ is a refinement of $\mathcal{G}$, which therefore has the same adherence point. But $\mathcal{G}$ is Cauchy and so convergent, by Proposition 8.27.

Proposition 12.5

Let $\{X_j\}$ be a family of complete uniform spaces. Then the uniform product $\prod X_j$ is also complete.

For let $\mathcal{F}$ be a Cauchy filter on the uniform product. Then $\pi_{j*} \mathcal{F}$ is a Cauchy filter on X_j for each index j, and so $\pi_{j*} \mathcal{F}$ converges to some point x_j of X_j since X_j is complete. Hence $\mathcal{F}$ converges to x, where $\pi_j(x) = x_j$ for each j, and so $\prod X_j$ is complete, as asserted.

Proposition 12.6

Let Y be a complete uniform space. Then for each set X the set Y^X of functions $X \to Y$ is uniformly complete.

By *uniformly complete* we mean, of course, complete in the uniformity of uniform convergence. To prove Proposition 12.6, let $\mathscr{F}$ be a uniformly Cauchy filter on Y^X. Then $\pi_{x*}\mathscr{F}$ is Cauchy in Y for each point x of X and so converges to a limit $\phi(x)$, say. In other words, $\mathscr{F}$ converges pointwise to the function $\phi \in Y^X$ thus defined and so, by Proposition 8.33, $\mathscr{F}$ converges uniformly to ϕ.

Now suppose that X is a topological space. Consider the subset Φ of Y^X consisting of *continuous* functions from X to Y, with the uniform topology. I assert that Φ is uniformly closed, i.e. closed in Y^X with the topology of uniform convergence. For let $\phi\colon X \to Y$ be a function which fails to be continuous at some point x of X. Then for some entourage D of Y the subset $\phi^{-1}(D[\phi(x)])$ of X contains no neighbourhood of x. If E is an entourage of Y such that $E \circ E \subset D$ then for each function $\psi \in \tilde{E}[\phi]$ we have that

$$\psi^{-1}(E[\phi(x)]) \subset \phi^{-1}(D[\phi(x)])$$

and so contains no neighbourhood of x. Then $\tilde{E}[\phi]$ is a neighbourhood of ϕ in Y^X consisting of functions which are discontinuous at x. Therefore Φ is closed in Y^X, as asserted. We deduce

Proposition 12.7

Let Y be a complete uniform space. Then for each topological space X the set Φ of continuous functions $X \to Y$ is uniformly complete.

For Y^X is uniformly complete, by Proposition 12.6, and so each uniformly closed subspace of Y^X is uniformly complete, by Proposition 12.2.

It is natural to ask what would happen if we replaced Cauchy filters by Cauchy sequences in the definition of completeness, as in

Definition 12.8

The uniform space X is sequentially complete if each Cauchy sequence of points of X is convergent.

For example, it is well known that the real line $\mathbb{R}$ is sequentially complete in the euclidean metric (this is either a theorem or an axiom depending on the method used to define the real number system).

For another example, let X be the Hilbert sequence space consisting of all sequences $x = \langle x_n \rangle$ of real numbers x_n such that $\sum x_n^2$ converges, with the metric given by

$$\rho(x, y) = \left(\sum (x_n - y_n)^2 \right)^{1/2}.$$

If $x^1, x^2, x^3, \ldots$ is a Cauchy sequence in X then, for each n, $(x_n^j)_{j=1}^{\infty}$ is a Cauchy sequence in the complete metric space $\mathbb{R}$ and thus converges to a point of $\mathbb{R}$, say x_n. Then if $x = \langle x_n \rangle$, the points $x - x^j$ eventually belong to X, hence $x = (x - x^j) + x^j$ must be in X, and so $\langle \rho(x, x^j) \rangle$ converges to 0. Thus X is sequentially complete.

Proposition 12.9

Let X be a complete uniform space. Then X is sequentially complete.

For let $\langle x_n \rangle$ be a Cauchy sequence in X. Then the associated elementary filter satisfies the Cauchy condition and so is convergent. Therefore $\langle x_n \rangle$ is convergent.

Proposition 12.10

The metric space X is sequentially complete if and only if X is complete, as a uniform space.

This means, of course, that the term sequentially complete is redundant where metric spaces are concerned.

Since complete always implies sequentially complete, as we have just seen, it is only necessary to show that the opposite implication holds in the metric case, as follows.

Suppose that the metric space X is sequentially complete. Let $\mathscr{F}$ be a Cauchy filter on X. Then for each positive ε there exists a member M of $\mathscr{F}$ which is ε-small, in the sense that $\rho(\xi, \eta) < \varepsilon$ whenever $\xi, \eta \in M$. So for each positive integer n we may set $\varepsilon = 1/n$ and obtain a nested sequence of members M_n of $\mathscr{F}$ which are $1/n$-small. Since each member of $\mathscr{F}$ is non-empty we may choose a point ξ_n of M_n and obtain a sequence $\langle \xi_n \rangle$ in X which satisfies the Cauchy condition and so is convergent. Let ξ be a limit point of the sequence. If U is a neighbourhood of ξ then $U_{1/n}(\xi) \subset U$ for some integer n. Choose an integer m such that $m > 2n$ and such that $\xi_m \in U_{1/2n}(\xi)$; then $\rho(\xi, \xi_m) < 1/2n$. Now for any point x of M_m we have $\rho(\xi_m, x) < 1/m$ since M_m is $1/m$-small, and so $\rho(\xi, x) < (1/2n) + (1/m) < 1/n$. Hence $x \in U$ and so $M_m \subset U$. Thus U is a member of $\mathscr{F}$ and so x is a limit point of $\mathscr{F}$ as required.

Proposition 12.11

The metric space X is complete if and only if there exists an accumulation point for each infinite totally bounded subset of X.

For suppose that X is complete, and let H be an infinite totally bounded subset of X. Since H is infinite we can choose a sequence of distinct points from H and then, since H is totally bounded, the sequence contains a Cauchy subsequence. The subsequence converges to some point x of X, since X is complete. But the subsequence is also composed of distinct points of H, and so x is an accumulation point of H.

To prove the converse, suppose that each infinite totally bounded subset of X admits an accumulation point. Let $\langle x_n \rangle$ be a Cauchy sequence in X. The underlying set H is totally bounded since for any $\varepsilon > 0$ there exists an integer k such that $\rho(x_m, x_n) < \varepsilon$ whenever $m, n \geq k$. Thus H is covered by the open balls $\{U_\varepsilon(x_n)\}$ for $n = 1, \ldots, k$; in other words, the set $\{x_n : n = 1, \ldots, k\}$ constitutes an ε-net. If H is finite, one of the terms of the sequence $\langle x_n \rangle$ must be repeated infinitely often, and the Cauchy sequence obviously converges to that point.

If H is infinite we can apply the hypothesis of Proposition 12.11 and obtain an accumulation point of H, say $x \in X$. We construct a subsequence $\langle x_{n(r)} \rangle$ of $\langle x_n \rangle$ which converges to x as follows. Since x is an accumulation point of H there exists an integer $n(1)$, say, such that $x_{n(1)} \in U_1(x) - \{x\}$. Suppose inductively that integers $n(1) < n(2) < \cdots < n(r)$ have been chosen so that $x_{n(i)} \in U_{1/i}(x) - \{x\}$ for $i = 1, 2, \ldots, r$. Put

$$\varepsilon = \min\left\{ \frac{1}{r+1}, \rho(x_1, x), \rho(x_2, x), \ldots, \rho(x_{n(r)}, x) \right\}.$$

Since $x_{n(i)} \neq x$ for all i we have $\varepsilon > 0$. Since x is an accumulation point of H there exists an integer $n(r+1)$ such that $x_{n(r+1)} \in U_\varepsilon(x) - \{x\}$. Now ε has been chosen to force $n(r+1) > n(r)$, since $\rho(x_i, x) \geq \varepsilon$ for all $i \leq n(r)$. Also $x_{n(r+1)} \in U_{1/r+1}(x)$. This completes the inductive step in the construction of the subsequence, which converges to x since $\rho(x_{n(r)}, x) < 1/r$ for all r.

Finally, we recall that $\langle x_n \rangle$ is a Cauchy sequence and so $\langle x_n \rangle$ itself converges to x, by Corollary 8.28. Thus X is complete, as asserted.

We return now to the general case and prove

Proposition 12.12

The uniform space X is complete if and only if the associated separated quotient space X' is complete.

For suppose that X is complete. If $\mathscr{F}'$ is a Cauchy filter on X' then the inverse image $\pi^*\mathscr{F}' = \mathscr{F}$ is a Cauchy filter on X, by Proposition 8.30, where $\pi\colon X \to X'$ is the natural projection. Now $\mathscr{F}$ converges to some point x of X, since X is complete, and so $\pi_*\mathscr{F}$ converges to $\pi(x)$ in X', since π is continuous. However, $\pi_*\pi^*\mathscr{F}' = \mathscr{F}'$, since π is surjective, and so $\mathscr{F}'$ is convergent. Thus X' is complete, which proves Proposition 12.12 in one direction.

Conversely, suppose that X' is complete. If $\mathscr{F}$ is a Cauchy filter on X then $\pi_*\mathscr{F}$ is a Cauchy filter on X', since π is uniformly continuous. Hence $\pi_*\mathscr{F}$ converges to some point x' of X', since X' is complete. Since the topology of X is induced from that of X', by Proposition 8.6, it follows that $\pi^*\pi_*\mathscr{F}$ converges to x for each point x of $\pi^{-1}(x')$. Hence the refinement $\mathscr{F}$ of $\pi^*\pi_*\mathscr{F}$ also converges to x, which completes the proof of Proposition 12.12.

Proposition 12.13

Let R be a compatible equivalence relation on the uniform space X, such that each of the equivalence classes $R[x]$ is complete. Suppose that the quotient uniform space X/R is complete. Then X is complete.

For let $\mathscr{F}$ be a Cauchy filter on X. Then $\pi_*\mathscr{F}$ converges to some point $\pi(x) \in X/R$. Consider the filter $\mathscr{G}$ generated by the subsets $D[M]$ of X, where D runs through the members of $\mathscr{F}$. If M is D-small then $D[M]$ is $(D \circ D \circ D)$-small and so $\mathscr{G}$ is a Cauchy filter on X, clearly coarser than $\mathscr{F}$. I assert that each member of $\mathscr{G}$ meets $R[x]$, so that the trace of $\mathscr{G}$ on $R[x]$ is defined. Assuming this we go on to argue that the trace converges to some point $\xi \in R[x]$, since $R[x]$ is complete. Then ξ is an adherence point, and hence a limit point, of $\mathscr{G}$ itself, since $\mathscr{G}$ is Cauchy. Therefore ξ is a limit point of the refinement $\mathscr{F}$ of $\mathscr{G}$, and so X is complete.

To prove the assertion we have to show that $D[M]$ meets $R[x]$ for each entourage D of X and member M of $\mathscr{F}$. By compatibility we have $R \circ E \subset D^{-1} \circ R$ for some entourage E of X. Since $\pi_*\mathscr{F}$ converges to $\pi(x)$ in X/R and since $\pi(E[x])$ is a neighbourhood of $\pi(x)$ we have that $\pi N \subset \pi(E[x])$ for some member $N \subset M$ of $\mathscr{F}$. Then

$$N \subset (R \circ E)[x] \subset D^{-1}[R[x]]$$

and so $D[N] \subset D[M]$ meets $R[x]$. This proves the assertion and completes the proof of Proposition 12.13. We deduce

Proposition 12.14

Let G be a topological group and let $K \subset H \subset G$ be subgroups. Suppose that G/H and H/K are complete in the right quotient uniformities. Then G/K is complete in the right quotient uniformity.

For since the natural projection from G to G/H is uniformly open it follows that the natural projection from G/K to G/H is uniformly open. Hence the equivalence relation whereby G/H is obtained from G/K is compatible and Proposition 12.14 follows.

Proposition 12.15

Let X be a uniform space and let A be a dense subset of X. Let $\phi: A \to Y$ be uniformly continuous, where Y is a complete separated uniform space. Then there exists a uniformly continuous extension $\psi: X \to Y$ of ϕ, and the extension is unique.

Uniqueness is an immediate consequence of Propositions 6.9 and 8.10.

To define ψ at a given point x of X we proceed as follows. The neighbourhood filter $\mathcal{N}_x$ is Cauchy and so the trace of $\mathcal{N}_x$ on A is a Cauchy filter on A, using Proposition 8.30. Since ϕ is uniformly continuous the direct image of the trace is a Cauchy filter $\mathcal{G}_x$ on Y. Since Y is complete $\mathcal{G}_x$ converges to a limit y in Y, and since Y is separated the limit is unique. We define $\psi(x) = y$.

Next we show that ψ, thus defined, coincides with ϕ on A. For suppose that $y = \phi(x)$, where $x \in A$. If V is a neighbourhood of y then $\phi^{-1}V$ is a neighbourhood of x in A, since ϕ is continuous, and so $\phi^{-1}V = U \cap A$ for some neighbourhood U of x in X. Then $\phi(U \cap A) \subset V$, and since $\phi(U \cap A) \in \mathcal{G}_x$ we have $V \in \mathcal{G}_x$. Thus $\mathcal{G}_x$ converges to y and so $\psi(x) = y = \phi(x)$.

To show that ψ is uniformly continuous, given an entourage E of Y, choose a symmetric entourage F of Y such that $F \circ F \circ F \subset E$. There exists an open entourage D of X such that $\phi \times \phi$ maps $D \cap (A \times A)$ into F. I assert that then $\psi \times \psi$ maps D into $F \circ F \circ F \subset E$. This will establish the uniform continuity of ψ on X.

So let x, x' be points of X such that $(x, x') \in D$. Let $\mathscr{F}, \mathscr{F}'$ be filters on A converging to x, x', respectively. Since D is open there exist members $M \in \mathscr{F}$, $M' \in \mathscr{F}'$ such that $M \times M' \subset D$ and so $\phi M \times \phi M' \subset F$. But since $\phi_* \mathscr{F}$ converges to $\psi(x)$, by definition, we have $\phi N \in F[\psi(x)]$ for some member N of $\mathscr{F}$. Similarly, $\phi N' \subset F[\psi(x')]$ for some member N' of $\mathscr{F}'$. Thus for some points $\xi \in M \cap N$ and $\xi' \in M' \cap N'$ we have that $(\phi(\xi), \phi(\xi'))$, $(\phi(\xi), \psi(x))$, and $(\phi(\xi'), \psi(x'))$ are contained in F, and so $(\psi(x), \psi(x')) \in F \circ F \circ F \subset E$, as required. This completes the proof of Proposition 12.15.

Corollary 12.16

Let X_1, X_2 be complete separated uniform spaces and let A_1, A_2 be dense subsets of X_1, X_2, respectively. If A_1 and A_2 are uniformly equivalent then X_1 and X_2 are uniformly equivalent.

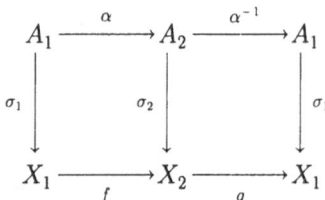

For let α, in the above diagram, be a uniform equivalence. By Proposition 12.15, $\sigma_2 \alpha$ extends to a uniformly continuous function f, as shown, while $\sigma_1 \alpha^{-1}$ extends to a uniformly continuous function g, as shown. Now the uniformly continuous function $gf: X_1 \to X_1$ is an extension of $\sigma_1 \alpha^{-1} \alpha = \sigma_1$. The identity id_{X_1} on X_1 is also such an extension, and so $gf = \mathrm{id}_{X_1}$ by uniqueness. Similarly, $fg = \mathrm{id}_{X_2}$. Therefore f is a uniform equivalence, as required.

In particular, suppose that X_1 and X_2 are metric spaces and that $\alpha: A_1 \to A_2$ is an isometry. Since each point of X_1 is the limit of a convergent sequence in A_1 it follows that the function $f: X_1 \to X_2$ we have defined is also an isometry. Thus we obtain

Proposition 12.17

Let X_1, X_2 be complete metric spaces and let A_1, A_2 be dense subsets of X_1, X_2, respectively. If A_1 and A_2 are isometrically equivalent then X_1 and X_2 are isometrically equivalent.

Proposition 12.18

Let X be a uniform space. Then X is compact if and only if X is complete and totally bounded.

For suppose that X is compact. Then each filter $\mathscr{F}$ on X admits an adherence point, by Corollary 5.10. If $\mathscr{F}$ is Cauchy then each adherence point is a limit point, by Proposition 8.27, which establishes completeness. In any case $\mathscr{F}$ can be refined by a filter $\mathscr{G}$ for which the adherence point is a limit point. Since $\mathscr{G}$ is then a Cauchy filter, by Proposition 8.26, this shows that X is also totally bounded.

Conversely, suppose that X is totally bounded. If $\mathscr{F}$ is a filter on X then there exists a Cauchy refinement $\mathscr{G}$ of $\mathscr{F}$, by Proposition 8.31. Further, suppose that X is complete. Then $\mathscr{G}$ admits a limit point, hence $\mathscr{F}$ admits an adherence point. Therefore X is compact, as asserted.

Note that the Heine–Borel theorem can be immediately deduced from this. For the unit interval I is obviously totally bounded, and complete by Proposition 12.2, therefore compact.

12.2 Metric completion

By a *metric completion* of a metric space X we mean a complete metric space $\hat{X}$ together with an isometric embedding of X as a dense subspace of $\hat{X}$. By Proposition 12.17 the metric completion of X, if it exists, is unique up to isometric equivalence, and so in this sense we may write *the* metric completion. For example, by Proposition 12.2 the metric completion of a subspace of a complete metric space is just the closure of the subspace. In fact, there are at least two general methods of constructing metric completions, as follows.

In the first method we consider the set $C^*(X)$ of bounded continuous real-valued functions on X. We denote the metric of X by ρ and the supremum metric of $C^*(X)$ by ρ^*. I assert that $C^*(X)$ is complete with this metric. For let $\langle \phi_n \rangle$ be a Cauchy sequence in $C^*(X)$. Then for each positive ε there exists an integer k such that

$$|\phi_m(x) - \phi_n(x)| \le \rho^*(\phi_m, \phi_n) < \varepsilon,$$

for all $m, n \ge k$ and all $x \in X$. Hence $\langle \phi_n(x) \rangle$, for each x, is a Cauchy sequence in $\mathbb{R}$ and so converges to some real number $\phi(x)$, since $\mathbb{R}$ is complete. Now for all $m, n \ge k$ and all $x \in X$ we have that $-\varepsilon < \phi_m(x) - \phi_n(x) < \varepsilon$ or, equivalently, that $\phi_n(x) - \varepsilon < \phi_m(x) < \phi_n(x) + \varepsilon$. Keeping n and x fixed while letting m tend to infinity we get $\phi_n(x) - \varepsilon \le \phi(x) \le \phi_n(x) + \varepsilon$ for all n and for all x. This implies that

$$\rho * (\phi_n, \phi) = \sup\{|\phi_n(x) - \phi(x)|: x \in X\} \le \varepsilon$$

for all $n \ge k$. Hence $\langle \phi_n \rangle \to \phi$, and since the convergence is uniform and ϕ_n is continuous and bounded for all n, we have that $\phi \in C^*(X)$. Thus $C^*(X)$ is complete, as asserted.

I now assert that X can be embedded isometrically as a subspace of $C^*(X)$. To see this, choose a point x_0 of X. For each point x of X consider the real-valued function $\phi(x): X \to \mathbb{R}$ defined by

$$\phi(x)(\xi) = \rho(\xi, x) - \rho(\xi, x_0) \qquad (\xi \in X).$$

Since the metric ρ is necessarily continuous, so is the function $\phi(x)$. Moreover, $\phi(x)$ is bounded since $|\phi(x)(\xi)| \leq \rho(x, x_0)$, by the triangle inequality, and so $\phi(x) \subset C^*(X)$. Now for $x, y \in X$ we have

$$\rho^*(\phi(x), \phi(y)) = \sup\{|[\phi(x)](\xi) - [\phi(y)](\xi)| : \xi \in X\}$$
$$= \sup\{|\rho(\xi, x) - \rho(\xi, y)| : \xi \in X\}.$$

For $\xi = y$ this yields

$$\rho^*(\phi(x), \phi(y)) \geq \rho(y, x) = \rho(x, y).$$

If $\rho(\xi, x) - \rho(\xi, y) > \rho(x, y)$ for some $\xi \in X$ then $\rho(\xi, x) > \rho(\xi, y) + \rho(x, y) \geq \rho(\xi, x)$ contrary to the triangle inequality. If $\rho(\xi, x) - \rho(\xi, y) < -\rho(x, y)$ for some $\xi \in X$ then $\rho(\xi, y) > \rho(x, y) + \rho(\xi, x)$ also contrary to the triangle inequality. We conclude that $|\rho(\xi, x) - \rho(\xi, y)| \leq \rho(x, y)$ for all $\xi \in X$, and hence that $\rho^*(\phi(x), \phi(y)) \leq \rho(x, y)$. Therefore

$$\rho^*(\phi(x), \phi(y)) = \rho(x, y),$$

for all $x, y \in X$. Thus ϕ embeds X isometrically into $C^*(X)$, and so X is isometrically equivalent to ϕX. Of course, ϕX is dense in $\text{Cl}\,(\phi X)$, and $\text{Cl}\,(\phi X)$ is complete by Proposition 12.2.

The second method of constructing the metric completion uses Cauchy sequences, generalizing the Cauchy procedure for constructing the real numbers out of the rationals. Let us say that the Cauchy sequences $\langle x_n \rangle$ and $\langle y_n \rangle$ in the metric space X are equivalent if the sequence $\langle \rho(x_n, y_n) \rangle$ of real numbers converges to zero. This equivalence relation partitions the collection of all Cauchy sequences in X into disjoint equivalence classes. We take these to be the points of a new space X^* having a metric ρ^* given by

$$\rho^*(x^*, y^*) = \lim_{n \to \infty} \rho(x_n, y_n),$$

where $\langle x_n \rangle, \langle y_n \rangle$ are representatives of $x^*, y^* \in X^*$, respectively. An isometric embedding $\phi: X \to X^*$ is defined so that $\phi(x)$, for each point x of X, is the equivalence class of Cauchy sequences having x as a limit point, in other words, the equivalence class containing the constant sequence at x. It is not difficult to show that X^* is complete and that ϕX is dense in X^*, in other words, X^* is a metric completion of X.

12.3 Uniform completion

We turn now to uniform spaces in general. By a *uniform completion* of a uniform space X we mean a complete uniform space $\hat{X}$ together with a uniform embedding of X as a dense subspace of $\hat{X}$. It is usual to require $\hat{X}$ also to be separated, and in that case Corollary 12.16 shows that the completion is unique up to uniform equivalence, and so in this sense we may write *the* separated completion. For example, by Proposition 12.2 the separated completion of a subspace of a complete separated uniform space is just the closure of the subspace. In case X is a metric space the metric completion is, *a fortiori*, the uniform completion. The first method of constructing the metric completion does not appear to extend to uniform spaces in general. However, the second method extends very satisfactorily, provided that Cauchy filters are used instead of Cauchy sequences. The details are as follows.

Let X be a uniform space. We consider first the set $\tilde{X}$ of Cauchy filters on X. It will clarify the exposition at this point if whenever a Cauchy filter $\mathscr{F}$ on X is regarded as a point of the set $\tilde{X}$ we denote that point by $\underline{\mathscr{F}}$. Let $\sigma \colon X \to \tilde{X}$ be the function which assigns to each point ξ of X the principal filter ε_ξ. Then σ is injective and we can make $\tilde{X}$ into a uniform space so that σ is a uniform embedding as follows.

Given a symmetric entourage D of X we denote by D^* the subset of $\tilde{X} \times \tilde{X}$ consisting of pairs $(\underline{\mathscr{F}}_1, \underline{\mathscr{F}}_2)$ such that the Cauchy filters $\mathscr{F}_1, \mathscr{F}_2$ on X have a D-small member in common. I claim that the family of subsets D^*, as D runs through the symmetric entourages of X, constitutes a base for a uniformity on $\tilde{X}$.

To show that the conditions for a uniformity base are satisfied we have to check four points. First, if $\mathscr{F}$ is a Cauchy filter on X then $(\underline{\mathscr{F}}, \underline{\mathscr{F}}) \in D^*$ for each symmetric entourage D of X; thus D^* contains the diagonal of $\tilde{X}$. Second, D^* is symmetric since D is symmetric. Third, if D' is a symmetric entourage of X such that $D' \circ D' \subset D$ then $D'^* \circ D'^* \subset D^*$; this is not quite so obvious, and can be seen as follows. Let $\mathscr{F}_1, \mathscr{F}_2, \mathscr{F}_3$ be Cauchy filters on X such that $(\underline{\mathscr{F}}_1, \underline{\mathscr{F}}_2) \in D'^*$ and $(\underline{\mathscr{F}}_2, \underline{\mathscr{F}}_3) \in D'^*$. Then $\mathscr{F}_1, \mathscr{F}_2$ have a common D'-small member M, say, and $\mathscr{F}_2, \mathscr{F}_3$ have a common D'-small member N, say. Both M and N are members of $\mathscr{F}_2$ and so M intersects N. Hence the union $M \cup N$ is $D' \circ D'$-small, therefore D-small. Since $M \cup N$ is a member of both $\mathscr{F}_1$ and $\mathscr{F}_3$ this shows that $(\underline{\mathscr{F}}_1, \underline{\mathscr{F}}_3) \in D^*$, as required.

For the last point which has to be checked let D_1, D_2 be symmetric entourages of X. Then the intersection $D_1 \cap D_2$ is also a symmetric entourage of X. Moreover, if $\mathscr{F}_1, \mathscr{F}_2$ are Cauchy filters on X such that $(\underline{\mathscr{F}}_1, \underline{\mathscr{F}}_2) \in D^*$, where $D = D_1 \cap D_2$, then $\mathscr{F}_1$ and $\mathscr{F}_2$ have a common D-small member M, say. Then M is both D_1-small and D_2-small, since $D \subset D_1$ and $D \subset D_2$, so that

$(\mathscr{F}_1, \mathscr{F}_2) \in D_1^* \cap D_2^*$. Thus $D^* \subset D_1^* \cap D_2^*$, and the last of the conditions for a uniformity base is satisfied.

It is clear from the construction that $\sigma: X \to \tilde{X}$ is a uniform embedding. We now show that $\tilde{X}$ is complete and that σX is dense in $\tilde{X}$. Let $\mathscr{F}$ be a Cauchy filter on X. Let D be a symmetric entourage of X. Then $\mathscr{F}$ contains a D-small member M, say. For each point x of M we have that $(\sigma(x), \mathscr{F}) \in D^*$, i.e. that $\sigma(x) \in D^*[\mathscr{F}]$. Therefore σX is dense in $\tilde{X}$. Furthermore, $\sigma M \subset D^*[\mathscr{F}]$ and so the principal filter of $\sigma(x)$ converges to the point $\mathscr{F}$ in $\tilde{X}$. Since X is uniformly equivalent to σX under σ the filters of the form $\sigma_*\mathscr{F}$ on $\tilde{X}$ are precisely the extensions of the Cauchy filters on σX and so, using Proposition 12.4, we conclude that $\tilde{X}$ is complete.

In general, the uniform space $\tilde{X}$ constructed in this way is not separated. One can, of course, always pass to the associated separated quotient space $\hat{X}$, say. The natural projection $\pi: \tilde{X} \to \hat{X}$ is uniformly continuous and so $\hat{X}$ is complete since $\tilde{X}$ is complete. Also $\pi\sigma X$ is dense in $\hat{X}$ since σX is dense in $\tilde{X}$ and π is a continuous surjection. In general $\pi\sigma$ is not injective. Suppose, however, that X itself is separated. Then for each pair ξ, η of distinct points of X there exists a symmetric entourage D of X which does not contain (ξ, η). The corresponding entourage D^* of $\tilde{X}$ does not contain $(\sigma(\xi), \sigma(\eta))$ and so $\pi\sigma(\xi) \neq \pi\sigma(\eta)$, by definition of $\hat{X}$. It follows that $\pi\sigma$ is a uniform embedding and so $\hat{X}$ satisfies all the requirements for a uniform completion of X.

Exercises

12.1. Show that the uniformity on the set $\mathbb{Z}$ of integers generated by the subsets

$$D_n = \{(\xi, \eta): \xi \equiv \eta \bmod n\},$$

for $n = 1, 2, \ldots$, is not complete.

12.2. Show that a subset Φ of Y^X, where X is a set and Y is a uniform space, is pointwise complete if (i) Φ is pointwise closed in Y^X and (ii) the closure of the projection $\pi_x \Phi$ is complete in Y for each point x of X.

12.3. Let X be a complete metric space with metric ρ. Let $\phi: X \to X$ be a function such that for some $k < 1$ the condition

$$\rho(\phi(\xi), \phi(\eta)) \cdot \rho(\xi, \eta)$$

is satisfied, for all $\xi, \eta \in X$. Show that ϕ is continuous and has precisely one fixed point.

12.4. Show that if $\{X_j\}$ is a family of uniform spaces then the uniform completion of the uniform product $\prod X_j$ is uniformly equivalent to the uniform product of the uniform completions $\{\hat{X}_j\}$.

12.5. Suppose that the topological group G admits a neighbourhood V of e which is complete for the left or right uniform structure. Show that G is complete.

12.6. Show that a compactly regular topological group is complete. Deduce that the rational line $\mathbb{Q}$ is not compactly regular.

12.7. Consider the real line $\mathbb{R}$ with the metric ρ given by
$$\rho(\xi, \eta) = |\xi(1 + |\xi|)^{-1} - \eta(1 + |\eta|)^{-1}| \qquad (\xi, \eta \in \mathbb{R}).$$
Show that $\mathbb{R}$ is not complete, with this metric, although the associated topology is euclidean.

12.8. Show that the interval $[0, 1)$ is complete with respect to the metric
$$\rho(\xi, \eta) = |(1 - \xi)^{-1} - (1 - \eta)^{-1}|.$$

12.9. Suppose that the metric ρ on the set X has the property that every closed and bounded set is compact. Show that X is complete.

12.10. Determine the completion of the real line $\mathbb{R}$ in the uniformity generated by the subsets
$$\Delta \mathbb{R} \cup ((\alpha, \infty) \times (\alpha, \infty))$$
for all real α.

Select Bibliography

1. M.A. Armstrong, *Basic Topology*, Springer-Verlag, New York, 1983.
2. N. Bourbaki, *Topologie Générale*, Hermann, Paris, 1980.
3. J. Dixmier, *General Topology*, Springer-Verlag, New York, 1984.
4. J. Dugundji, *Topology*, Allyn and Bacon, Boston, 1965.
5. R. Engelking, *Outline of General Topology*, North-Holland, Amsterdam, 1968.
6. S.A. Gaal, *Point-Set Topology*, Academic Press, New York, 1964.
7. J.R. Isbell, *Uniform Spaces*, Amer. Math. Soc., Providence, 1964.
8. I.M. James, *General Topology and Homotopy Theory*, Springer-Verlag, New York, 1984.
9. I.M. James, *Introduction to Uniform Spaces*, Cambridge University Press, 1990.
10. J.L. Kelley, *General Topology*, van Nostrand, New York, 1955.
11. J. Munkres, *Topology: A First Course*, Prentice-Hall, Englewood Cliffs, NJ, 1974.
12 J.-I. Nagata, *Modern General Topology*, North-Holland, Amsterdam, 1985.
13. W. Page, *Topological and Uniform Structures*, Wiley, New York, 1978.
14. W. Roelcke and S. Dierholf, *Uniform Structures on Topological Groups and their Quotients*, McGraw-Hill, New York, 1981.
15. L.A. Steen and J.A. Seebach, *Counterexamples in Topology*, Holt, Rinehart and Winston, New York, 1970.
16. S. Willard, *General Topology*, Addison-Wesley, Reading, MA, 1968.

Solutions to Exercises

Chapter 1

1.1 Clearly $H - K = H \cap (X - K)$. Hence, in the first case, $H - K$ is the intersection of two open sets and, in the second, the intersection of two closed sets.

1.2. If $x \in U \cap \mathrm{Cl}\, H$ and W is an open neighbourhood of x, then $W \cap U$ is also an open neighbourhood of x. Since $x \in \mathrm{Cl}\, H$, by Proposition 1.10 $W \cap U \cap H \neq \emptyset$. The result follows by Proposition 1.10 again.

1.3. Clearly $\mathrm{Int}(\mathrm{Cl}\, U) \subset \mathrm{Cl}\, U$ and hence, $\mathrm{Cl}(\mathrm{Int}(\mathrm{Cl}\, U)) \subset \mathrm{Cl}(\mathrm{Cl}\, U) = \mathrm{Cl}\, U$. For the reverse inclusion, if $x \in \mathrm{Cl}\, U$ and W is an open neighbourhood of x, then $U \cap W \neq \emptyset$. Now, since U is open, $U \subset \mathrm{Int}(\mathrm{Cl}\, U)$, thus $W \cap \mathrm{Int}(\mathrm{Cl}\, U) \neq \emptyset$. Hence, $x \in \mathrm{Cl}(\mathrm{Int}(\mathrm{Cl}\, U))$.

1.4. If U is regularly open then $A = X - U$ is closed and $\mathrm{Cl}(\mathrm{Int}\, A) \subset \mathrm{Cl}\, A = A$. If $x \in A$ and W is an open neighbourhood of x, then since $x \notin U = \mathrm{Int}(\mathrm{Cl}\, U)$, $W \cap (X - \mathrm{Cl}\, U) \neq \emptyset$. Thus $x \in \mathrm{Cl}(X - \mathrm{Cl}\, U) = \mathrm{Cl}(\mathrm{Int}(X - U)) = \mathrm{Cl}(\mathrm{Int}\, A)$. Thus $A = \mathrm{Cl}(\mathrm{Int}\, A)$ and A is regularly closed.

Now, if $B = \mathrm{Int}(\mathrm{Cl}\, A)$, then B is open and $B \subset \mathrm{Cl}\, B$, so $B \subset \mathrm{Int}(\mathrm{Cl}\, B)$. Also, $\mathrm{Cl}\, B \subset \mathrm{Cl}(\mathrm{Cl}\, A) = \mathrm{Cl}\, A$ and thus $\mathrm{Int}(\mathrm{Cl}\, B) \subset \mathrm{Int}(\mathrm{Cl}\, A) = B$. Thus $B = \mathrm{Int}(\mathrm{Cl}\, B)$ and B is regularly open. The proofs of the other two results are similar.

1.5 $\mathbb{R} - \{0\}$ is open but not regularly open and $\{0\}$ is closed but not regularly closed.

1.6 In general $\mathrm{Cl}(\bigcup H_j) \neq \bigcup(\mathrm{Cl}\, H_j)$ (for example, consider $H_n = (\frac{1}{n}, 1]$ in $\mathbb{R}$ for $n = 1, 2, 3, \ldots$), however equality holds for finite families. We prove that $\mathrm{Cl}(H_1 \cup H_2) = (\mathrm{Cl}\, H_1) \cup (\mathrm{Cl}\, H_2)$. The result then follows by induction. Easily, $\mathrm{Cl}\, H_i \subset \mathrm{Cl}(H_1 \cup H_2)$ for $i = 1, 2$. Also, $(\mathrm{Cl}\, H_1) \cup (\mathrm{Cl}\, H_2)$ is closed and contains $H_1 \cup H_2$ and the reverse inclusion then follows.

$\mathrm{Cl}(\bigcap H_j) \neq \bigcap (\mathrm{Cl}\, H_j)$, even for finite families (for example $H_1 = [0, 1)$ and $H_2 = (1, 2]$ in $\mathbb{R}$).

For interiors, $\mathrm{Int}([0, 1] \cup (1, 2]) \neq \mathrm{Int}\,[0, 1] \cup \mathrm{Int}\,(1, 2]$. For finite families $\mathrm{Int}(\bigcap H_j) = \bigcap(\mathrm{Int}\, H_j)$ (proof as before) but this equality fails in general (for example $H_n = (-\frac{1}{n}, \frac{1}{n})$ in $\mathbb{R}$, for $n = 1, 2, 3, \ldots$).

1.7. Yes. If $x \in X$, then $X - \{x\}$ is not dense and hence $\mathrm{Cl}(X - \{x\}) = X - \{x\}$. Thus, $X - \{x\}$ is closed and $\{x\}$ is open.

1.8. No. Consider the cocountable topology on an uncountable set.

1.9. Easily $\varnothing, X \in \mathcal{T}$. If $U, V \in \mathcal{T}$, then either $x_0 \notin U \cap V$ or $X - (U \cap V) = (X - U) \cup (X - V)$, a union of two finite sets. If $U_\lambda \in \mathcal{T}$ for all $\lambda \in \Lambda$, then either $x_0 \notin \bigcup_\lambda U_\lambda$ or $x_0 \in U_{\lambda_0}$ for some λ_0 and then $X - \bigcup_\lambda U_\lambda \subset X - U_{\lambda_0}$ which is finite. Thus $\mathcal{T}$ is a topology on X. If $x \neq x_0$ then $x_0 \notin \{x\}$, so $\{x\}$ is open, and $X - \{x\}$ is cofinite so $\{x\}$ is closed also.

1.10. Since ϕ is an injection, $\rho(\xi, \eta) = 0$ if and only if $\phi(\xi) = \phi(\eta)$ if and only if $\xi = \eta$. Also ρ is clearly symmetric and the triangle inequality for ρ follows from the triangle inequality in $\mathbb{R}$.

1.11. Pick $A \subset \mathbb{R}$ such that $\varnothing \neq A \neq \mathbb{R}$ and define a topology $\mathcal{T}$ on $\mathbb{R}$ by $\mathcal{T} = \{\varnothing, A, \mathbb{R} - A, \mathbb{R}\}$.

Chapter 2

2.1. Let $A_j = \{t \in \mathbb{R}: p_j(t) = 0\} = p_j^{-1}(0)$ which is closed since $\{0\}$ is closed in $\mathbb{R}$ and p_j is continuous. The required set is $\bigcap_j A_j$ which is therefore closed.

2.2. The *open* sets in $E \times E$ are $\varnothing$, $\{(0,0)\}$, $\{(0,0), (0,1)\}$, $\{(0,0), (1,0)\}$, $\{(0,0), (0,1), (1,0)\}$ and $E \times E$. The closed sets are then the complements of these 6 sets in $E \times E$.

2.3. For each $j \in \mathbb{N}$, let $D_j = \{0, 1\}$ with the discrete topology. Let $X = \prod_{j \in \mathbb{N}} D_j$ and let $x \in X$ be given by $\pi_j(x) = 0$ for all j. Then $\{x\}$ is not open in X because otherwise there is some non-empty restricted product neighbourhood $U = \prod_{j \in \mathbb{N}} U_j \subset \{x\}$. Assume $U_j = D_j$ for all $j \notin \{j_1, \ldots, j_n\}$, and choose $y \in X$ such that $\pi_j(y) = 0$ for $j = j_1, \ldots, j_n$ and $\pi_j(y) = 1$ otherwise. Thus $y \in U - \{x\}$ which is a contradiction.

2.4.   If u denotes the inversion map on the product and $\prod U_j$ is a product open
       set, then $u^{-1} \prod U_j = \prod u_j^{-1} U_j$ where u_j denotes the inversion map on G_j.
       Since each u_j is continuous this is also a product open set. For multiplica-
       tion, $m^{-1} \prod U_j = \prod m_j^{-1} U_j$ (with, as usual, the obvious rearrangement
       of factors) which is a product open set. For the box topology we thus
       deduce that u and m are continuous. For the product topology, the
       same results hold for restricted product open sets since if $U_j = G_j$, then
       $u_j^{-1} U_j = G_j$ and similarly for m_j.

2.5.   Let $d: G \times G \to G$, denote the division map, that is $d(g, h) = gh^{-1}$. Then
       d is continuous and $\Delta G = d^{-1}(e)$ is thus closed.

2.6.   If $D \subset \mathbb{R}$ is a closed subgroup and $d \in D$, then $\{nd: n \in \mathbb{Z}\} \subset D$. So, if for
       all $\varepsilon > 0$ there is $d \in (0, \varepsilon) \cap D$, then D is dense in $\mathbb{R}$ and thus equals $\mathbb{R}$
       since it is closed. Otherwise, there is a smallest strictly positive $\alpha \in D$
       (the infimum is attained since D is closed). In this case $D = \alpha \mathbb{Z}$. Other-
       wise, there is some $d \in D - \alpha \mathbb{Z}$ with $d > 0$ and then $\alpha n < d < \alpha(n+1)$
       for some n and $0 < d - \alpha n < \alpha$ which is a contradiction since
       $d - \alpha n \in D$.

2.7.   If f is a non-constant continuous function on X, then for some $x, y \in X$,
       $f(x) \neq f(y)$. Choose $\varepsilon > 0$ such that $U_\varepsilon(f(x)) \cap U_\varepsilon(f(y)) = \varnothing$. Then
       $f^{-1} U_\varepsilon(f(x))$ and $f^{-1} U_\varepsilon(f(y))$ are disjoint, non-empty and open in X.
       Since X is infinite and has the cofinite topology, any two non-empty
       open sets must intersect which is a contradiction.

2.8.   If U is a neighbourhood of e, then there is an open neighbourhood V of e
       such that $e \in V \subset U$. Since inversion is a homeomorphism, V^{-1} is also an
       open neighbourhood of e. Then $V \cap V^{-1}$ is symmetric, open and
       $e \in V \cap V^{-1} \subset U$.

2.9   By (i) the family $\{gU: U \in \Gamma\}$ is a filter base for a filter $\mathcal{N}_g$. We check
       that $\{\mathcal{N}_g: g \in G\}$ satisfies the conditions in Definition 1.20. For coher-
       ence, if $N \in \mathcal{N}_g$ choose $gU \subset N$. If $g' = gu \in gU$, then we claim that
       $gU \in \mathcal{N}_{g'}$. If $h = g'v$ for some $v \in U$, then $h = guv \in gU$ and thus
       $g'U \subset gU$ and so $gU \in \mathcal{N}_{g'}$. Thus we have the required neighbourhood
       base. We need to check it is a topological group. If $U \in \Gamma$ and $g \in G$,
       then choose V as in (ii). Then, if $v \in V$, $(g^{-1}v)^{-1} = gg^{-1}v^{-1}g \in gU$.
       Thus if u denotes the inversion map, $u(g^{-1}V) \subset gU$ and u is continuous.
       Also if $v_1, v_2 \in W \subset U \cap V$, with $W \in \Gamma$, then for each $h \in G$, $hv_1gv_2 =$
       $hgg^{-1}v_1gv_2 \in hgU$ and thus $m(hW \times gW) \subset hgU$ and m is continuous.
       It is an exercise in group theory that the families in parts (a)–(c) satisfy
       conditions (i) and (ii).

2.10. For each $N \in \mathcal{N}(e)$, we have $N^{-1} \in \mathcal{N}(e)$ and so, if $g \in \mathrm{Cl}\, E$, then $N^{-1}g \cap E \neq \varnothing$ and thus $g \in N \cdot E$. Conversely, take an arbitrary neighbourhood Ng of g (with $N \in \mathcal{N}(e)$). By assumption $g \in N^{-1} \cdot E$ and therefore $g = n^{-1}f$ for some $n \in N$ and $f \in E$. Thus $f \in E \cap Ng$ and hence $g \in \mathrm{Cl}\, E$.

2.11. Clearly, $\bigcup \{V : V \text{ open}, V^2 \subset U\} \subset \{g : g^2 \in U\}$. For the reverse inclusion, if $g^2 \in U$ then by continuity of multiplication, there is an open set $V \ni g$ such that $m(V \times V) \subset U$, that is $V^2 \subset U$.

2.12. By Example 2.14 we only need show H is open. Assume $g \in \mathrm{Int}\, H$. Then for some open neighbourhood U of e, $gU \subset H$. If $h \in H$ then $hU = hg^{-1}gU \subset hg^{-1}H = H$ and thus $hU \subset H$ for all $h \in H$ and H is open.

2.13. By Example 2.15, $\mathrm{Cl}\, K$ is a subgroup. The map $\phi : G \times G \to G$ given by $\phi(h, k) = h^{-1}kh$, is continuous. Since K is normal in H, $\phi(H \times K) \subset K$ and by Proposition 2.3, $\phi(\mathrm{Cl}\, H \times \mathrm{Cl}\, K) \subset \mathrm{Cl}\, K$. But since H is dense this means that $\mathrm{Cl}\, K$ is normal in G.

2.14. By basic group theory $Z(G)$, the centre of G is a normal subgroup. Assume $k \in \mathrm{Cl}\, Z(G) - Z(G)$, so there is $h \in G$ such that $kh \neq hk$. The map ϕ which sends g to $g^{-1}h^{-1}gh$ is continuous and by assumption $\phi(k) \neq e$ and thus $k \notin \phi^{-1}(e)$ which is closed. So there is an open $U \ni k$ such that $U \cap \phi^{-1}(e) = \varnothing$. But $U \cap Z(G) \neq \varnothing$ (since $k \in \mathrm{Cl}\, Z(G)$) which is a contradiction since $Z(G) \subset \phi^{-1}(e)$.

2.15. (i) If V is the intersection of all open subgroups then it is clearly a subgroup. Assume $g \in V$, $h \in G$ and U is an open subgroup. Since conjugation is a homeomorphism, hUh^{-1} is an open subgroup and thus $g \in hUh^{-1}$. So $h^{-1}gh \in U$ and since U was arbitrary, $h^{-1}gh \in V$ and V is normal. The proof of (ii) is similar.

Chapter 3

3.1.   If U is open and $\varnothing \neq U \neq X$, then choose $x \in U$ and $y \notin U$. Then $\{x\} = U \cap \{x, y\}$ is open in $\{x, y\}$, a contradiction. The corresponding assertion for the discrete topology is not true. A counterexample is the real line with the euclidean topology.

3.2.   From the assumptions, $\{\mathrm{Int}\, A_n : n = 2, 3, \ldots\}$ is an open covering of X. For each n, $\phi | A_n$ is continuous and thus so is $\phi | \mathrm{Int}\, A_n$. The result follows from Corollary 3.10.

3.3.   Since H is a subgroup, $\text{Cl } H$ is itself a topological group. If $\text{Cl } H - H$ is not dense in $\text{Cl } H$, then there is a non-empty open set U in $\text{Cl } H$ such that $U \subset H$. Thus H has non-empty interior in $\text{Cl } H$ and then, by Exercise 2.12, H is closed in $\text{Cl } H$. Hence there is a closed set F in G such that $F \cap \text{Cl } H = H$ and so H is closed in G which is a contradiction.

3.4.   Assume $U \cap \text{Cl } \{e\}$ is non-empty and open in $\text{Cl } \{e\}$ and take any $g \in \text{Cl } \{e\}$. By Exercise 2.8 and continuity of multiplication, there is a symmetric open $V \ni e$ such that $V^2 \subset U$ (note $e \in U$ since $U \cap \text{Cl } \{e\} \neq \varnothing$). By Exercise 2.10, we have $\text{Cl } V \subset V^2$ and thus $g \in \text{Cl } \{e\} \subset \text{Cl } V \subset U$. Thus $U \cap \text{Cl } \{e\} = \text{Cl } \{e\}$.

3.5.   (i) Clearly ϕ is continuous and defining $\psi: \mathbb{R} \to \mathbb{R} \times \mathbb{R}$ by $\psi(x) = (x, 0)$ gives a continuous right inverse for ϕ. The result follows from Proposition 3.14. (ii) Similarly $\psi: [0, \infty) \to \mathbb{R} \times \mathbb{R}$, defined by $\psi(x) = (\sqrt{x}, 0)$ gives a continuous right inverse for ϕ.

3.6.   Define $f: (X \times Y)/R \to Y$ by $f(R[(\xi, \eta)]) = \eta$. By definition of R, f is well-defined and an injection. It is clearly a surjection. If U is open in Y and $\pi: (X \times Y) \to (X \times Y)/R$ is the quotient map, then $\pi^{-1}(f^{-1}U) = X \times U$ which is open in $X \times Y$ and so $f^{-1}U$ is open in $(X \times Y)/R$. Thus f is continuous. If U is open in $(X \times Y)/R$, then $U = \{R[(\xi, \eta)]: \xi \in X, \eta \in U' \subset Y\}$ for some U' and $\pi^{-1}U = X \times U'$ so U' is open. But $fU = U'$ so f has a continuous inverse.

3.7. $I/R = \{\mathbb{Q} \cap I, I - \mathbb{Q}\}$. If U is open in I/R and $\mathbb{Q} \cap I \in U$, then $\mathbb{Q} \cap I \subset \pi^{-1}U$ which is open in I. Since $I - \mathbb{Q}$ is dense in I, there is some $p \in (\pi^{-1}U) \cap (I - \mathbb{Q})$ and hence $I - \mathbb{Q} \in U$. A similar argument shows that no open set contains $I - \mathbb{Q}$ but not $I \cap \mathbb{Q}$.

3.8. $S^1/Z_2 = \{Z_2[e^{i\theta}]: \theta \in [0, \pi)\}$. Define $\phi: S^1/Z_2 \to S^1$, by $\phi(Z_2[e^{i\theta}]) = e^{2i\theta}$ and $\psi: S^1 \to S^1/Z_2$ by $\psi(e^{i\theta}) = Z_2[e^{i\theta/2}]$. Then, ϕ is a continuous bijection with continuous inverse ψ.

3.9.   It is easy to see that $\phi(x, y, z) = \phi(-x, -y, -z)$. Thus $\psi: S^2/Z_2 \to \mathbb{R}^4$ given by $\psi(Z_2[x]) = \phi(x)$ is well-defined. It is also an injection since $\phi(x) = \phi(y)$ if and only if, either $x = y$, or x and y are antipodal points. Since $\phi = \psi\pi$ it is also continuous by Proposition 3.12. If (a, b, c, d) is in the image of ψ then it is easy to see that it is the image of the equivalence class of $(\sqrt{a}, \frac{b}{\sqrt{a}}, \frac{d\sqrt{a}}{b})$. Thus ψ has a continuous left inverse and is therefore an embedding.

Chapter 4

4.1.   Assume $\{Y_j\}$ is an open covering and thus $\{\phi^{-1}Y_j\}$ is a covering of X. If
each of the restrictions is open and U is open in X, then $\phi U =$
$\bigcup_j \phi(U \cap \phi^{-1}Y_j)$, a union of sets each of which is open in its respective Y_j
and thus in Y. If each of the restrictions is closed and A is closed in X then
$Y - \phi A = \bigcup_j (Y_j - \phi(A \cap \phi^{-1}Y_j))$ which is, similarly, a union of open
sets too. Now assume $\{Y_j\}$ is a finite closed covering. If each of the restrictions is open and U is open in X, then $Y - \phi U = \bigcup_j (Y_j - \phi(U \cap \phi^{-1}Y_j))$,
which is a finite union of sets each of which is closed in its respective Y_j
and thus in Y. Hence ϕU is open. If each of the restrictions is closed
and A is closed in X then $\phi A = \bigcup_j \phi(A \cap \phi^{-1}Y_j)$ which is, similarly, a
finite union of closed sets.

4.2.   If $r < 1$, then $\beta^{-1}(r, 1] = \{y \in Y : \exists\, x \in X \text{ such that } \alpha(x) > r \text{ and}$
$x \in \phi^{-1}(y)\} = \phi(\alpha^{-1}(r, 1])$ which is open since ϕ is open and α is continuous. If $s > 0$ then $\beta^{-1}[0, s) = \bigcup_{t < s} \{y \in Y : \phi^{-1}(y) \subset \alpha^{-1}[0, t)\} =$
$\bigcup_{t < s} \{Y - \phi(\alpha^{-1}[t, 1])\}$ which is open since ϕ is closed and α is continuous. Thus $\beta^{-1}(r, s) = \beta^{-1}(r, 1] \cap \beta^{-1}[0, s)$ is open for all $r < 1$ and $s > 0$
and β is continuous.

4.3.   The given set is a hyperbola and is closed in $\mathbb{R}^2$. Let B be its image under
addition. Clearly $B \subset \mathbb{R} - \{0\}$ but since the line $x = -y$ is an asymptote
to the hyperbola, for all $\varepsilon > 0$ there is (ξ, η) on the hyperbola such that
$\xi + \eta < \varepsilon$. Thus 0 is an adherence point of B and B is not closed.

4.4.   The given set, A say, is closed since it is the inverse image of $\{1\}$ under the
continuous map that sends (x, y) to $xy(x + y)$. Now for $(\xi, \eta) \in A$,
$\xi\eta = \frac{1}{\xi+\eta}$ and thus 0 is not in the image of A under multiplication.
However, for $M > 0$, if $N = \frac{-M^2 + \sqrt{M^4 + 4M}}{2M}$, then $(M, N) \in A$ and
$MN = \frac{1}{M+N} = (\frac{M}{2} + \frac{\sqrt{M^4 + 4M}}{2M})^{-1}$ which can be chosen to be as close to 0 as
possible by choosing M large enough. Thus 0 is an adherence point of the
image of A and the image is not closed.

4.5.   Choose $x_0 \in H \cap K$. For $y, z \in H \cup K$, either $y, z \in H$ and thus $d(y, z) \leq$
diam H,   or $y, z \in K$ and   thus $d(y, z) \leq$ diam K or   otherwise
$d(y, z) \leq d(y, x_0) + d(x_0, z) \leq$ diam $H +$ diam K.

4.6.   Assume A is open and U is open in X. If $U \cap A = \emptyset$ then $\pi^{-1}(\pi U) = U$,
otherwise, $\pi^{-1}(\pi U) = U \cup A$. In either case $\pi^{-1}(\pi U)$ is open in X and
thus πU is open in X/A. If A is closed and C is closed in X, then, similarly,
$\pi^{-1}(\pi C)$ is either C or $C \cup A$ and thus πC is closed in X/A.

Chapter 5

5.1. Since X is not compact it is either not closed or not bounded.

(i) If X is not bounded, let ϕ be the identity map on X. If X is not closed, choose $a \in \mathrm{Cl}\, X - X$ and define $\phi(x) = \frac{1}{x-a}$.

(ii) If X is not bounded define $\phi(x) = \frac{1}{1+|x|}$ which is bounded by 0 and 1. If X is bounded but not closed then choose $a \in \mathrm{Cl}\, X - X$ and define $\phi(x) = |x - a|$ which is bounded since diam $X = $ diam $\mathrm{Cl}\, X$. In either case 0 is the infimum of ϕX but is not attained.

5.2. Fix $F_0 \in \mathscr{F}$. Assume for a contradiction that no intersection of a finite subfamily of $\mathscr{F}$ is contained in U. Let $\mathscr{G} = \{F_0 \cap (F - U) \colon F \in \mathscr{F}\}$, then $\mathscr{G}$ is a filter base for a filter on F_0 which by compactness has an adherence point x. Since elements of $\mathscr{G}$ are closed in F_0, $x \in F - U$ for all $F \in \mathscr{F}$ which is a contradiction.

5.3. Fix $e \in E$. For each $f \in F$ choose open U_f and V_f such that $(e, f) \in U_f \times V_f \subset W$. Then $\{V_f\}$ is an open covering of F and so there is a finite subcovering $\{V_{f_1}, \ldots, V_{f_n}\}$. Let $U_e = \bigcap U_{f_i}$ and $V_e = \bigcup V_{f_i}$ which are both open. Thus for each $e \in E$ we have $\{e\} \times F \subset U_e \times V_e \subset W$. Now $\{U_e\}$ is an open covering of E, and so there is a finite subcovering $\{U_{e_1}, \ldots, U_{e_m}\}$ of E. Letting $U = \bigcup U_{e_i}$ and $V = \bigcap V_{e_i}$ gives the required result.

5.4. If $\mathscr{U}$ is an open covering of X^+, then $* \in U_0$ for some $U_0 \in \mathscr{U}$. By definition of the topology, $A = X^+ - U_0$ is a compact subspace of X. The trace of $\mathscr{U}$ on A is an open covering of A and thus there are $U_1, \ldots, U_n \in \mathscr{U}$ which cover A. Thus $\{U_0, U_1, \ldots, U_n\}$ is the required subcovering.

5.5. If ϕ is proper then taking T to be a single point shows that ϕ is a closed map. If $y \in Y$, then $g \colon \phi^{-1}(y) \times T \to \{y\} \times T$ defined by $g = (\phi \times \mathrm{id}_T)|\phi^{-1}(y) \times T$ is easily seen to be closed. If $\pi \colon \{y\} \times T \to T$ is the second projection, then π is closed. Hence $\pi g \colon \phi^{-1}(y) \times T \to T$ is closed. But πg is the second projection and so $\phi^{-1}(y)$ is compact by definition.

For the converse we shall use Proposition 4.15. Assume $(y, t) \in Y \times T$ and W is open in $X \times T$ and contains $(\phi \times \mathrm{id}_T)^{-1}((y, t)) = \phi^{-1}(y) \times \{t\}$. By (half of) the argument in Exercise 5.3, there are open U and V such that $\phi^{-1}(y) \times \{t\} \subset U \times V \subset W$. Since ϕ is closed, by Proposition 4.15, there is an open set V' in Y containing y such that $\phi^{-1}V' \subset U$ and then $(\phi \times \mathrm{id}_T)^{-1}(V' \times V)$ is a subset of W and by Proposition 4.15 again, $\phi \times \mathrm{id}_T$ is closed.

5.6. $A \cdot B$ is the continuous image under multiplication of $A \times B$ which is compact by Proposition 5.3.

5.7. Take $x \in G - C \cdot E$, then $C \cap xE^{-1} = \emptyset$. Now E is closed and therefore xE^{-1} is closed. For each $y \in C$ there is an open neighbourhood V_y of e such that $V_y V_y y \cap xE^{-1} = \emptyset$. By compactness, C is covered by finitely many $V_y y$ for $y \in C$. Let V be the intersection of these finitely many V_y. Then $e \in V$ and $VC \cap xE^{-1} = \emptyset$. Thus $V^{-1}x \cap C \cdot E = \emptyset$ and we have proved that $G - C \cdot E$ is open.

5.8. Since H is open, the cosets of H are open and give a disjoint open covering of G. Since G is compact there must be a finite subcovering but by disjointness the covering has no proper subcovering and thus must have been finite to begin with.

5.9. We first show that for a fixed $y \in C$ there is a symmetric open neighbourhood V_y of e such that $xV_y x^{-1} \subset U$ whenever $x \in V_y y$. Choose symmetric open neighbourhoods of e, V_1 and V_2 such that $V_1^3 \subset U$ and $yV_2 y^{-1} \subset V_1$ and let $V_y = V_1 \cap V_2$. If $x \in V_y y$, then $xV_y x^{-1} \subset xV_2 x^{-1} = (xy^{-1})yV_2 y^{-1}(yx^{-1}) \subset V_1^3$. Now by compactness C is covered by finitely many $V_y y$. Let V be the corresponding finite intersection of the V_y. The result then follows.

5.10. If A is closed in G, then $\pi A = \{gH : g \in A\}$. Thus $\pi^{-1}(\pi A) = \{k : kH = gH \text{ for some } g \in A\} = \{k : kg^{-1} \in H \text{ for some } g \in A\} = H \cdot A$ which is closed by Exercise 5.7. Hence πA is closed.

Chapter 6

6.1. If $R[\xi] \neq R[\eta]$, then without loss of generality, there exists $x \in \text{Cl}\{\xi\} - \text{Cl}\{\eta\}$. So there is an open set U containing x such that $\eta \notin U$ and, since $x \in \text{Cl}\{\xi\}$, $\xi \in U$. If $y \in \pi^{-1}(\pi U)$ then $\text{Cl}\{y\} = \text{Cl}\{z\}$ for some $z \in U$ and thus $z \in \text{Cl}\{y\}$ and hence $y \in U$. So $\pi^{-1}(\pi U) = U$ which is open and thus πU is an open neighbourhood of $R[\xi]$ which does not contain $R[\eta]$.

6.2. Clearly $H \subset \bigcap\{N : N \text{ is a neighbourhood of } H\}$. For the reverse inclusion, if $x \notin H$ then, by T_1, $X - \{x\}$ is an open neighbourhood of H and so $x \notin \bigcap\{N : N \text{ is a neighbourhood of } H\}$.

6.3. The result follows immediately from Proposition 6.2.

6.4. Take any $S[x] \in X/S$. Then $\pi^{-1}S[x]$ is $\{x\}$ or $\{-x, x\}$, which in either case is closed in X, so $\{S[x]\}$ is closed and X/S is T_1. If U and V are open in X/S and $S[-1] \in U$ and $S[1] \in V$, then $-1 \in \pi^{-1}U$ and

$1 \in \pi^{-1}V$ and both these sets are open in X. Thus, there is some δ with $0 < \delta < 1$ such that $-1 + \delta \in U$ and $1 - \delta \in V$. Thus, $S[1 - \delta] \in U \cap V$ and X/S is not Hausdorff.

6.5. Clearly $A \subset \bigcap\{B: B$ is a closed neighbourhood of $A\}$. For the reverse inclusion, if $x \notin A$ then, by regularity, there exist disjoint open U and V such that $x \in U$ and $A \subset V$. Thus $X - U$ is a closed neighbourhood of A which does not contain x.

6.6. It is easy to show that $\phi^{-1}C = \pi_X((X \times C) \cap \Gamma_\phi)$ where π_X is the projection from $X \times Y$ onto X. So if C is closed in X the two cases are as follows:

(i) $(X \times C) \cap \Gamma_\phi$ is a closed subspace of Γ_ϕ and is hence compact. Thus $\phi^{-1}C$ is compact (since π_X is continuous) and hence it is closed (since X is Hausdorff).

(ii) Since Y is compact, the projection π_X is closed. Now, $(X \times C) \cap \Gamma_\phi$ is closed in $X \times Y$, so $\phi^{-1}C$ is closed.

6.7. By normality there exist disjoint open U_1 and V_1 such that $E \subset U_1$ and $F \subset V_1$. Now E and $X - U_1$ are disjoint and closed so by normality again there exist disjoint open U and U' such that $E \subset U$ and $X - U_1 \subset U'$. Thus $E \subset U \subset \text{Cl } U \subset U_1$. Similarly there is an open V such that $F \subset V \subset \text{Cl } V \subset V_1$. Then $\text{Cl } U$ and $\text{Cl } V$ are as required.

6.8. By assumption there is a symmetric open neighbourhood of e such that $V \cap H = \{e\}$. If $x \in \text{Cl } H$, then $xV \cap H \neq \emptyset$. If $y \in xV \cap H$, then $x \in yV$ and $\{y\} = y(V \cap H) = yV \cap H$ is closed. However $x \in \text{Cl }(yV \cap H)$ by Exercise 1.2, and so $x \in H$.

6.9. We use the alternative notation for products, that is $x(j)$ instead of $\pi_j(x)$. Define $\phi: (\prod X_j)' \to \prod X_j'$ by $\phi[x](j) = [x(j)]$. Now $[x] = [y]$ if and only if $x \in \text{Cl}\{y\}$ if and only if for each j, $x(j) \in \text{Cl}\{y(j)\}$ if and only if $[x(j)] = [y(j)]$ for each j. Hence ϕ is well-defined and an injection. It is clearly a surjection. If $\prod U_j'$ is a restricted product open set in $\prod X_j'$ then $\pi^{-1}(\phi^{-1} \prod U_j') = \prod \pi_j^{-1} U_j'$ where π is the quotient map on $\prod X_j$ and π_j is the quotient map on X_j. This last set is open and thus ϕ is continuous. Similarly if U is open in $(\prod X_j)'$, then $\pi^{-1}U$ is the union of restricted product open sets $\prod U_j$ in $\prod X_j$. Then ϕU is the union of sets of the form $\prod \pi_j U_j$ which are open since each π_j is open.

6.10. Every subspace is compact, and therefore closed, since X is Hausdorff.

6.11. Write $X = \mathbb{R}$ with new topology. Clearly one-point sets are closed so X is T_1. Assume $x \neq y$ and U and V are open with $x \in U$ and $y \in V$. So both $X - U$ and $X - V$ are bounded and hence there exist $N, M > 0$ such that

$X - U \subset [-N, N]$ and $X - V \subset [-M, M]$. If $p > \max(N, M)$, then $p \in U \cap V$ and so X is not Hausdorff.

Chapter 7

7.1.   Clearly $\Delta \subset U_\varepsilon = \{(\xi, \eta): |\xi - \eta| \in \mathbb{Q}, |\xi - \eta| < \varepsilon\}$ and each U_ε is symmetric. It is easy to check that $U_{\varepsilon/2} \circ U_{\varepsilon/2} \subset U_\varepsilon$ since if $x - y \in \mathbb{Q}$ and $y - z \in \mathbb{Q}$, then $x - z \in \mathbb{Q}$. Thus $\{U_\varepsilon: \varepsilon > 0\}$ generates a uniformity.

7.2.   If R is a base for $\mathscr{U}$, then $R = \bigcap \mathscr{U}$ and thus R is an equivalence relation (page 108). Conversely, if R is an equivalence relation, then by reflexivity $\Delta \subset R$ and thus $R \neq \varnothing$ and $\{R\}$ is a filter base. Since, also $R = R^{-1}$ and $R \circ R = R$, the conditions in Definition 7.1 are satisfied.

7.3.   If $\delta > 0$ then by uniform continuity of ϕ there is an entourage V of X such that $V \subset (\phi \times \phi)^{-1} U_{\delta\varepsilon^2}$. Now, if $(x, y) \in V$, then

$$\left| \frac{1}{\phi(x)} - \frac{1}{\phi(y)} \right| = \frac{|\phi(y) - \phi(x)|}{|\phi(x)\phi(y)|} \leq \frac{|\phi(y) - \phi(x)|}{\varepsilon^2} < \delta.$$

Thus $V \subset (\frac{1}{\phi} \times \frac{1}{\phi})^{-1} U_\delta$ and $1/\phi$ is uniformly continuous.

7.4.   Let $\varepsilon = \frac{1}{2}$. For each positive $\delta < 1$ let $x = \frac{1}{\delta}$ and $y = \frac{1}{\delta} + \frac{\delta}{2}$. Then $|x - y| < \delta$ but $|e^x - e^y| = e^{\frac{1}{\delta}}(e^{\frac{\delta}{2}} - 1) \geq \frac{1}{\delta}(\frac{\delta}{2}) = \frac{1}{2}$.

7.5.   Consider the space X used in Exercise 2.3. Then Δ is not an entourage otherwise there is a restricted product open neighbourhood in X^2, $\prod V_j \subset \Delta$. But for some j, $V_j = \{0, 1\}^2$ and so there is $(x, y) \in \Delta$ such that $\pi_j(x) \neq \pi_j(y)$ which is a contradiction.

7.6.   Assume A and B are bounded. Given an entourage D there exist integers n and m, and finite subsets S and T of A and B respectively such that $A \subset D^n[S]$ and $B \subset D^m[T]$. If $p = \max(n, m)$, then we obtain $A \cup B \subset D^p[S] \cup D^p[T] \subset D^p[S \cup T]$. Thus $A \cup B$ is bounded.

7.7.   Take any entourage $D \cap (A \times A)$ of A where D is an entourage of X. Since R is compatible there is an entourage D' such that $R \circ D' \subset D \circ R$. For if $S \circ (D' \cap (A \times A)) \subset (D \cap (A \times A)) \circ S$ and thus S is compatible. Now $(a, b) \in S \circ (D' \cap (A \times A))$, then $(a, b) \in R \circ D' \subset D \circ R$ so there is a c such that $(a, c) \in R$ and $(c, b) \in D$. But $a \in A$ so, by saturation, $c \in A$ and, since $b \in A$, we have $(a, c) \in S$ and $(c, b) \in D \cap (A \times A)$ as required.

7.8.   Clearly ρ is symmetric and positive and $\rho(\xi, \eta) = 0$ if and only if $\xi = \eta$. Also $\rho(\xi, \eta) + \rho(\eta, \zeta)$ represents the length of a path on the surface of the sphere going from ξ to η and then from η to ζ and so this must be at least as big as the length of the shortest path from ξ to ζ, that is $\rho(\xi, \zeta)$.

If d denotes the euclidean metric then obviously $U_\varepsilon^\rho \subset U_\varepsilon^d$. Conversely, let C be the great circle containing ξ and η and assume the arc of this circle from ξ to η subtends an angle ε (which is thus $\rho(\xi, \eta)$). If $a = d(\xi, \eta)$ then by the cosine rule $a^2 = 2 - 2\cos\varepsilon$. So if $\delta = \sqrt{2 - 2\cos\varepsilon}$ then $U_\delta^d \subset U_\varepsilon^\rho$. Thus the two metrics are equivalent.

7.9.   Let $V_{n,\varepsilon} = \{(\xi, \eta): \rho_n(\xi, \eta) < \varepsilon\}$, then the uniformity determined by ρ_n is generated by the $V_{n,\varepsilon}$ for $\varepsilon > 0$. If $n > m$ we show that there does not exist $\delta > 0$ such that $V_{m,\delta} \subset V_{n,1}$. The required result then follows. So for $\delta > 0$ choose $\xi > 0$ such that $\xi^{n-m} > \frac{2}{\delta}$ and choose $\eta < \xi$ such that $\xi^m = \eta^m + \frac{\delta}{2}$. So $(\xi, \eta) \in V_{m,\delta}$ but $|\xi^n - \eta^n| = \xi^n - \eta^n \geq \xi^{n-m}(\xi^m - \eta^m) \geq \frac{2}{\delta}\frac{\delta}{2} = 1$.

7.10.   Choose $m > 0$ such that $|\phi(x)|, |\psi(x)| < m$ for all $x \in X$. Given $\varepsilon > 0$, since ϕ and ψ are uniformly continuous, there are entourages V_1 and V_2 such that $V_1 \subset (\phi \times \phi)^{-1}(U_{\frac{\varepsilon}{2m}})$ and $V_2 \subset (\psi \times \psi)^{-1}(U_{\frac{\varepsilon}{2m}})$. Choose an entourage $V \subset V_1 \cap V_2$. If $(x, y) \in V$ then

$$|\phi(x)\psi(x) - \phi(y)\psi(y)|$$
$$\leq |\phi(x)\psi(x) - \phi(y)\psi(x)| + |\phi(y)\psi(x) - \phi(y)\psi(y)|$$
$$= |\psi(x)||\phi(x) - \phi(y)| + |\phi(y)||\psi(x) - \psi(y)|$$
$$\leq m\frac{\varepsilon}{2m} + m\frac{\varepsilon}{2m} = \varepsilon.$$

Thus $V \subset (\phi \cdot \psi \times \phi \cdot \psi)^{-1} U_\varepsilon$.

7.11.   (i) If E is an entourage of Y then $D = (\phi \times \phi)^{-1}E$ is an entourage of X by uniform continuity. Since $\psi\phi$ is uniformly open there is an entourage F of Z such that $F[\psi\phi(x)] \subset \psi\phi(D[x])$ for each $x \in X$. If $y \in Y$, then, since ϕ is a surjection, $y = \phi(x)$ for some $x \in X$. Then $F[\psi(y)] \subset \psi(\phi(D[x])) \subset \psi(E[y])$ and ψ is uniformly open.

(ii) If D is an entourage of X then, since $\psi\phi$ is uniformly open, there is an entourage F of Z such that $F[\psi\phi(x)] \subset \psi\phi(D[x])$ for each $x \in X$. Now $E = (\psi \times \psi)^{-1}F$ is an entourage of Y by uniform continuity. If $y \in E[\phi(x)]$ then $\psi(y) \in F[\psi\phi(x)] \subset \psi(\phi(D[x]))$. Since ψ is an injection, $y \in \phi(D[x])$.

Chapter 8

8.1.   If D is an entourage, then $\bigcup_{j=1}^n H_j \times H_j \subset D$ where $\{H_j\}$ is a finite partition of X. If $\mathscr{U}$ is an ultrafilter, then $H_j \in \mathscr{U}$ for some j by Proposition 1.18, and H_j is thus D-small. Hence $\mathscr{U}$ is Cauchy.

If X is infinite there is a non-principal ultrafilter on X (page 17), which by the above is Cauchy. The uniformity is thus not discrete (page 138).

8.2. Let $U_\varepsilon = \{(x, y): |x - y| < \varepsilon\}$. The new uniformity is generated by sets of the form $V_\delta = \{(x, y): |x^3 - y^3| < \delta\}$. Elementary differential calculus shows that the points on the curves $y^3 = x^3 + \delta$ and $x^3 = y^3 + \delta$ where the perpendicular distance to the diagonal is maximal occur on the anti-diagonal. At these points the perpendicular distance is $\sqrt{2}\,\sqrt[3]{\delta/2}$. For the curves $y = x + \varepsilon$ and $x = y + \varepsilon$ the perpendicular distance from the diagonal is constantly $\varepsilon/\sqrt{2}$. Therefore if $\delta \leq \varepsilon^3/4$, then $V_\delta \subset U_\varepsilon$. Thus the new uniformity is a refinement of the euclidean uniformity. It is a strict refinement by Exercise 7.9. Hence euclidean entourages are also entourages in the new uniformity and thus Cauchy sequences in the new uniformity are Cauchy in the euclidean uniformity. For the converse, assume $\langle x_n \rangle$ is a Cauchy sequence in the euclidean uniformity and hence bounded ($|x_n| \leq M$ for all n say). Then, given $\varepsilon > 0$ there is a K such that $|x_n - x_m| < \varepsilon/3M^2$ for all $m, n \geq K$. Thus

$$|x_n^3 - x_m^3| = |x_n - x_m||x_n^2 + x_n x_m + x_m^2| \leq 3M^2\varepsilon/3M^2 = \varepsilon$$

(and hence $(x_n, x_m) \in V_\varepsilon$) for all $m, n \geq K$.

8.3. The given family is easily a filter base since $\mathscr{F}$ is a filter. Conditions (i) and (ii) of Definition 7.1 are clearly satisfied. For a given $M \in \mathscr{F}$, it is easy to check that if $(x, y), (y, z) \in (M \times M) \cup \Delta$, then $(x, z) \in (M \times M) \cup \Delta$ and thus condition (iii) holds too. If $x \in X$, then either $\{x\} \in \mathscr{F}$ and hence $D = \Delta$ is an entourage, or $X - \{x\} \in \mathscr{F}$ and then $D = ((X - \{x\}) \times (X - \{x\})) \cup \Delta$ is an entourage. In either case, $D[x] = \{x\}$ and the uniform topology is discrete.

8.4. For each $x \in X$, there exists $V_x \in \Gamma$ such that $x \in V_x$. Since the open neighbourhoods of ΔX form a base for the uniformity given by Proposition 8.20, there is some open neighbourhood of ΔX, E_x such that $E_x[x] \subset V_x$. Choose symmetric open neighbourhoods of ΔX, D_x such that $D_x \circ D_x \subset E_x$ for each x. Then $\{D_x[x]\}$ is an open covering of X and there is a finite subcovering given by $D_{x_1}, \ldots, D_{x_n}$. Let $D = \bigcap D_{x_i}$. Then for each $x \in X$, $x \in D_{x_i}[x_i]$ for some i and

$$D[x] \subset (D_{x_i} \circ D_{x_i})[x_i] \subset E_{x_i}[x_i] \subset V_{x_i} \in \Gamma.$$

8.5. If $y \notin D[A]$ then for each $x \in A$, $(x, y) \notin D$. Therefore, there is an open $U_x \ni x$ and $V_x \ni y$ such that $(U_x \times V_x) \cap D = \varnothing$. Hence $\{U_x : x \in A\}$ covers A and so there is a finite subcovering $\{U_{x_1}, \ldots, U_{x_n}\}$. If we let $V = \bigcap V_{x_i}$, then $y \in V \subset X - D[A]$.

8.6. Let $B = \bigcup_{n \in \mathbb{N}} D^n[A]$. If $x \in B$, then for some n there is $y \in A$ such that $(y, x) \in D^n$. If $z \in D[x]$ then $(x, z) \in D$ and so $z \in D^{n+1}[y] \subset B$. Thus $x \in D[x] \subset B$ and B is open. Choose E to be a symmetric entourage such that $E \subset D$. By Corollary 8.12, $\mathrm{Cl}\, B \subset E[B] \subset D[B] \subset B$ and thus B is closed.

8.7. It is sufficient to prove that if G is a topological group and H is a dense subgroup on which the left and right uniformities coincide then the left and right uniformities coincide on G. We show that for each neighbourhood V of e, there is a neighbourhood W of e such that $g^{-1}Wg \subset V$ for all $g \in G$.

By regularity choose neighbourhoods of e, V_1 and V_2 such that $\mathrm{Cl}\, V_1 \subset V_2 \subset \mathrm{Cl}\, V_2 \subset V$. Since $V_1 \cap H$ is a neighbourhood of e in H and the left and right uniformities coincide, there is a neighbourhood T of e in H such that $g^{-1}Tg \subset V_1 \cap H$ for all $g \in H$. Let $W = \mathrm{Int}\,(\mathrm{Cl}\, T)$. By denseness of H, $e \in W$. We claim that $g^{-1}wg \in V$ for all $g \in G$ and $w \in W$. For $g \in H$ the map ϕ which sends x to $g^{-1}xg$ is continuous and from above $\phi T \subset V_1$. By continuity $\phi W \subset \phi\, \mathrm{Cl}\, T \subset \mathrm{Cl}\, V_1 \subset V_2$. Thus $g^{-1}wg \in V_2$ for all $g \in H$ and $w \in W$. Now for a fixed $w \in W$ the map ψ which takes x to $x^{-1}wx$ is continuous and we have just shown that $\psi H \subset V_2$. By continuity $\psi G = \psi\, \mathrm{Cl}\, H \subset \mathrm{Cl}\, V_2 \subset V$, and our claim is proved.

8.8. A base for the neighbourhood filter of the neutral element consists of the subsets $W_S = \{\theta \in G \colon \theta(t) = t\ \forall t \in S\}$, where S runs through the finite subsets of $\mathbb{Z}$. As before (see page 143) the left relation L_S determined by W_S consists of pairs (ϕ, ψ) of elements of G such that $\phi(t) = \psi(t)$ for all $t \in S$. A similar argument shows that the right relation R_S consists of pairs such that $\phi^{-1}(t) = \psi^{-1}(t)$ for all $t \in S$. So let $S = \{0\}$ and let T be any finite subset of $\mathbb{Z}$. Choose distinct, non-zero integers t_1, t_2, t_3 not in T and define $\phi, \psi \in G$ as follows: $\phi(0) = t_1$, $\phi(t_1) = t_2$, $\phi(t_2) = t_3$, $\phi(t_3) = 0$ and $\psi(0) = t_2$, $\psi(t_2) = t_1$, $\psi(t_1) = t_3$, $\psi(t_3) = 0$, all other points are fixed. Then $(\phi, \psi) \in R_T - L_S$. Thus we have shown that the right uniformity does not refine the left uniformity. A similar argument shows that the left is not a refinement of the right.

Chapter 9

9.1. Note that $X_1 \cap X_2 \neq \varnothing$ otherwise X_1 and X_2 are proper, non-empty closed and open subsets of X, contradicting connectedness of X. If $f \colon X_1 \to \mathbb{D}$ is continuous where $\mathbb{D}$ is discrete, then $f|X_1 \cap X_2$ is also continuous and is thus constant (taking value d say). If $f(x) = e \neq d$

for some $x \in X_1$ then $f^{-1}(e) \cap X_2 = \emptyset$ and so $X - f^{-1}(e) = X_2 \cup (X_1 - f^{-1}(e))$. Therefore $f^{-1}(e)$ is open in X. However, $f^{-1}(e)$ is closed in X_1 and thus closed in X which is a contradiction. Thus f is constant and X_1 is connected. Similarly X_2 is connected.

9.2. The function $f : L \to \{0, 1\}$, given by $f(x) = 0$ if $x < 0$ and $f(x) = 1$ if $x \geq 0$ is continuous when $\{0, 1\}$ is given the discrete topology.

9.3. The set in question is connected because it is pathwise-connected. There are various cases but for instance, if (x, y) and (x', y') are both in the set and x and x' are both irrational, then they can be joined by a path in the set consisting of straight lines via the points (x, x') and (x', x'). The other cases are similar. The result remains true (with analogous proofs) if irrational is replaced by rational.

9.4. The set is pathwise-connected. For instance, if $(x, y), (x', y') \in X$ where x is rational and x' irrational, then a path consisting of straight lines via $(x, 0)$ and $(x', 0)$ will do.

9.5. The result is proved by induction on n. The case $n = 2$ follows from Proposition 9.12. If $\bigcup_{i=1}^{n-1} A_i$ is connected, then since A_n is connected and $A_n \cap (\bigcup_{i=1}^{n-1} A_i) \neq \emptyset$, then $\bigcup_{i=1}^{n} A_i$ is connected by Proposition 9.12 again.

9.6. If $\phi^{-1} Y'$ is connected then $Y' = \phi(\phi^{-1} Y')$ is connected by Proposition 9.5. Conversely, note that, since the restriction of a quotient map to a closed or open set is also a quotient map, it is enough to prove that if Y is connected then so is X. So assume Y is connected and C is a non-empty closed and open set in X. If $y \in \pi C$, then $\pi^{-1}(y) \cap C \neq \emptyset$ and thus $\pi^{-1}(y) \subset C$ since $\pi^{-1}(y)$ is connected. Thus $C = \pi^{-1}(\pi C)$ and we can deduce that $\pi(C)$ is closed and open in Y. Since Y is connected, $\pi C = Y$ and hence $C = \pi^{-1}(\pi C) = X$.

9.7. If X is totally disconnected and $\{y_1, y_2\} \subset Y \subset X$, then $\{y_1, y_2\}$, as a subspace of X, is disconnected and therefore is a disconnected subspace of Y.

If $C \subset \prod X_\lambda$ where each X_λ is totally disconnected, and C is connected, then for each λ, π_λ is continuous and therefore, $\pi_\lambda C$ is connected and is thus a one-point set. Hence, C is also a one-point set.

9.8. The set $A = \bigcup_{n \in \mathbb{N}} U^n$ is a subgroup and since U is open, each U^n is open and thus A is an open subgroup. By Example 2.14 A is also closed and hence $A = G$ since G is connected. Finally, apply Example 2.14 with

$H = \text{Cl}\{e\}$.

9.9.   Let $a \in H$ and let U be an open neighbourhood of a such that $U \cap H = \{a\}$. The map which sends g to $g^{-1}ag$ is continuous and so there is a symmetric open neighbourhood V of the neutral element such that $v^{-1}av \in U$ for all $v \in V$. Since H is normal, this implies that $v^{-1}av \in U \cap H$ for all $v \in V$, that is $v^{-1}av = a$ for all $v \in V$. By Exercise 9.8, each $x \in G$ is in some V^n, that is $x = v_1 v_2 \ldots v_n$ (for some $v_i \in V$) and thus $x^{-1}ax = a$ and H is central.

9.10.   By Theorem 5.9, since $\mathscr{B}$ consists of closed sets, the intersection $D = \bigcap \mathscr{B}$ is non-empty (it contains the adherence point) and is clearly closed. If D is disconnected there are disjoint proper closed and open subsets K_1 and K_2 of D such that $D = K_1 \cup K_2$. Since D is closed K_1 and K_2 are closed in X and by normality (Corollary 6.27) there are disjoint open W_1 and W_2 such that $K_i \subset W_i$. By Exercise 5.2, there is a $B \in \mathscr{B}$ such that $B \subset W_1 \cup W_2$. Then $W_1 \cap B$ is non-empty (it contains $D \cap K_1$) and open in B and its complement in B is $W_2 \cap B$ which is also non-empty and open in B which contradicts B being connected.

9.11.   For some open V in X, $V \cap H = C$. For each $c \in C$ there is a connected, open (in X) subset E_c such that $c \in E_c \subset V$. Let $U = \bigcup_{c \in C} E_c$. Then U is open, $H \cap U = C$ and U is connected by Corollary 9.13.

9.12.   By applying Exercise 2.9 we can show that it is a topological group ((ii) holds since $\mathbb{Q}$ is commutative). Suppose that $q \in \mathbb{Q} - U_t$ for $t \in \mathbb{Z}$, then we claim that $(q + U_t) \cap U_t = \varnothing$. If not then $q + \frac{m'p^t}{n'} = \frac{mp^t}{n}$ for some n, n' not divisible by p. Then $q = \frac{(mn' - nm')p^t}{nn'}$ which is a contradiction since nn' is not divisible by p. Thus each U_t is closed (and open). Since $0 \in U_t$ for each t, the component of 0 must be contained in U_t for each t. But $\bigcap_t U_t = \{0\}$ and so the component of 0 is $\{0\}$. Since it is a topological group all components are thus one-point sets as required.

9.13.   Suppose that C is a connected subspace of X which contains more than one point. So $\frac{1}{n} \in C$ for some n. Now $\{\frac{1}{n}\}$ is closed in C (which is T_1) and $\{\frac{1}{n}\} = (\frac{1}{n+1}, \frac{1}{n-1}) \cap C$ and so $\{\frac{1}{n}\}$ is also open in C which is a contradiction. Hence C is a one-point set.

9.14.   Since ϕ is uniformly open and $X \times X$ is an entourage, there is an entourage E of Y such that $E[\phi(x)] \subset \phi((X \times X)[x]) = \phi X$ for each $x \in X$. Thus, by induction, we can deduce that $E^n[\phi(X)] \subset \phi X$ for all n. So if $y \in Y$ and $x \in X$, then since Y is uniformly connected, there is an n

such that $(\phi(x), y) \in E^n$ by Proposition 9.35 and so $y \in \phi X$.

9.15. We use Proposition 9.35. Let D be a symmetric entourage of X so that $(\pi \times \pi)D = E$ is an entourage of X/R. If $\xi, \eta \in X$ then $(\pi(\xi), \pi(\eta)) \in E^n$ for some n. Thus there are $x_0, \ldots, x_n \in X$ such that $(\xi, x_0), (x_n, \eta) \in R$ and $(x_i, x_{i+1}) \in R \circ D \circ R$ for $i = 0, 1, \ldots, n-1$. We claim that, for each $i \leq n-1$, $(x_i, x_{i+1}) \in D^m$ for some m. Now for some a, b we have $(a, b) \in D$, $(x_i, a) \in R$, and $(b, x_{i+1}) \in R$. Since each equivalence class is uniformly connected $(x_i, a) \in D^j$ and $(b, x_{i+1}) \in D^k$ for some j and k. Thus $(x_i, x_{i+1}) \in D^{j+k+1}$. Putting all this together $(\xi, \eta) \in D^r$ for some r as required.

9.16. It is easy to check that the defined relation is an equivalence relation. Suppose that $y \notin [x]$ (the equivalence class of x). Then there is an entourage D such that for each n, $(x, y) \notin D^n$. We claim that $D^{-1}[y] \cap [x] = \varnothing$ and thus $[x]$ is closed. If $z \in D^{-1}[y] \cap [x]$ then $(z, y) \in D$ and $(x, z) \in D^m$ for some m and thus $(x, y) \in D^{m+1}$, a contradiction.

Let X be $\{(0,0)\} \cup \{(1,0)\} \cup \bigcup_{n \in \mathbb{N}}([0,1] \times \{1/n\})$ with the euclidean topology. Two points x, y are related by the given equivalence relation if for each $\varepsilon > 0$ there is a finite set of points $x_0, \ldots, x_n$ in X such that $x = x_0$, $x_n = y$ and $|x_i - x_{i+1}| < \varepsilon$ for $0 \leq i \leq n-1$. The equivalence class containing $(0,0)$ is thus the set $\{(0,0), (1,0)\}$ which is discrete and therefore not uniformly connected.

Chapter 10

10.1. Assume X is Hausdorff and the sequence $\langle x_n \rangle$ converges to x. If $y \neq x$ then there are disjoint open sets U and V such that $x \in U$ and $y \in V$. Since $x_n \to x$ there is an N such that $x_n \in U$ for all $n \geq N$ and so $x_n \notin V$ for all $n \geq N$. Thus y is not a limit of the sequence. Conversely assume every sequence has a unique limit. If X is not Hausdorff then there exist $x \neq y$ such that every neighbourhood of x has non-empty intersection with every neighbourhood of y. Since X is first countable the neighbourhood filters at x and y have countable bases $(U_n)_{n \in \mathbb{N}}$ and $(V_n)_{n \in \mathbb{N}}$ respectively. Without loss of generality we may assume that these bases are decreasing. Choose $x_n \in U_n \cap V_n$ for each n. Then $\langle x_n \rangle$ is a sequence converging to both x and y which is a contradiction.

10.2. If $\{U_\alpha\}$ is an uncountable family of pairwise disjoint non-empty open sets and D is dense, then for each α we may choose $x_\alpha \in D \cap U_\alpha$. Now by disjointness these are distinct points, so D is uncountable and X is not separable.

10.3. Assume that Y is an uncountable subset of X with no accumulation point. Then for each $x \in X$, x is not an accumulation point and there is an open $U_x \ni x$ such that $U_x \cap Y \subset \{x\}$. Since X is Lindelöf, the open covering $\{U_x\}$ has a countable subcovering $\{U_{x_1}, U_{x_2}, \ldots\}$. Since Y is uncountable there is a $y \in Y$ such that $y \neq x_i$ for each $i \in \mathbb{N}$ and thus by construction $y \notin U_{x_i}$ for each i, which is a contradiction to $\{U_{x_1}, U_{x_2}, \ldots\}$ being a covering.

10.4. It is easy to show that the set of points (x, y) with x and y both rational is dense in the Sorgenfrey plane. The antidiagonal subspace is uncountable and discrete (see page 52) and thus is not separable (cf. the comment preceding Proposition 10.30).

10.5 Consider the projection $\pi_Y \colon X \times Y \to Y$ which is closed by compactness of X. For each $y \in Y$, $\pi_Y^{-1}(y)$ is homeomorphic to X and is hence compact. So if $\{U_\lambda\}$ is an open covering of $X \times Y$, then for each $y \in Y$ there is a finite set Λ_y such that $\pi_Y^{-1}(y) \subset \bigcup \{U_\lambda \colon \lambda \in \Lambda_y\}$. Let $V_y = \bigcup \{U_\lambda \colon \lambda \in \Lambda_y\}$ and let $W_y = Y - \pi_Y((X \times Y) - V_y)$. Then $y \in W_y$ for all y so $\{W_y\}$ is an open covering of Y. Since Y is Lindelöf, there is a countable subcovering $\{W_{y_1}, W_{y_2}, \ldots\}$ of Y. One can then check that $\{U_\lambda \colon \lambda \in \bigcup_{i=1}^\infty \Lambda_{y_i}\}$ is the required countable subcovering.

10.6. Assume that $\{U_\lambda\}$ is an open covering of the Sorgenfrey line L and let V_λ be the interior of U_λ in the euclidean topology. Let $F = L - \bigcup_\lambda V_\lambda$. If $x \in F$ then there is $\lambda(x)$ and $r(x) > x$ such that $[x, r(x)) \subset U_{\lambda(x)}$. By definition of F, if $x \neq x'$ then $[x, r(x)) \cap [x', r(x')) = \varnothing$. So F is countable otherwise we have uncountably many non-empty pairwise disjoint open intervals $(x, r(x))$ in $\mathbb{R}$ which contradicts Exercise 10.2. By Proposition 10.3, $L - F$ with the euclidean topology is second countable and $\{V_\lambda\}$ covers $L - F$. By Proposition 10.8 there is a countable subcovering $\{V_{\lambda_i}\}$ of $L - F$ and thus $\{U_{\lambda_i} \colon i \in \mathbb{N}\} \cup \{U_{\lambda(x)} \colon x \in F\}$ is a countable subcovering of $\{U_\lambda\}$.

By Exercise 10.3, $L \times L$ is not Lindelöf since the antidiagonal is closed and discrete and therefore has no accumulation point but is uncountable.

10.7. As in Exercise 6.6(ii), it is enough to prove that the projection π_X is closed. So assume F is closed in $X \times Y$ and $\langle x_n \rangle$ is a sequence in $\pi_X F$ with $x_n \to x$. Choose $y_n \in Y$ with $(x_n, y_n) \in F$. Either there is $y \in Y$ such that $y = y_n$ for infinitely many n or there is an accumulation point y of $\{y_n\}$ (Proposition 10.14). In either case $(x, y) \in \mathrm{Cl}\, F = F$ and thus $x \in \pi_X F$ which is therefore closed.

10.8. Assume that X is countably compact and A is a closed discrete subspace.

A has no accumulation point since, for $x \notin A$, $X - A$ is an open neighbourhood of x disjoint from A and for $x \in A$ there is an open U such that $U \cap A = \{x\}$. By Proposition 10.14 A is finite. Conversely, let A be an infinite subset of X. If A is not closed then there exists $x \in \mathrm{Cl}\, A - A$. If A is closed but not discrete then there is $x \in A$ such that $U \cap A \neq \{x\}$ for all open $U \ni x$. In either case x is an accumulation point of A and X is countably compact by Proposition 10.14.

10.9.   Necessity follows from Proposition 7.25. Assume every countable subset of X is totally bounded and D is an entourage of X. If $D[S] \neq X$ for all finite S, then, as in the proof of Proposition 10.29 construct $x_i \in X$ for $i \in \mathbb{N}$, such that $(x_i, x_j) \notin D$ whenever $i \neq j$. If $H = \{x_1, x_2, \ldots\}$, then H is totally bounded so $(D \cap (H \times H))[S] = H$ for some finite $S \subset H$ which is a contradiction.

Chapter 11

11.1.   For each $y \in F$ there is a continuous $f_y \colon X \to I$ such that $f_y(x) = 0$ if $x \in E$ and $f_y(y) = 1$. Thus $\{f_y^{-1}(\frac{1}{2}, 1] : y \in F\}$ is an open covering of F from which by compactness we can extract a finite subcovering $\{f_{y_1}^{-1}(\frac{1}{2}, 1], \ldots, f_{y_n}^{-1}(\frac{1}{2}, 1]\}$. Defining $g = \max(f_{y_i})$ we have g is continuous, $g(x) = 0$ for all $x \in E$ and $F \subset g^{-1}(\frac{1}{2}, 1]$. Now define $\alpha \colon X \to I$ by $\alpha(x) = 2\min(\frac{1}{2}, g(x))$.

11.2.   Assume that X is compact and let $F_\alpha = \alpha^{-1}(0)$ for each $\alpha \in \Gamma$. Each F_α is closed and non-empty and any finite intersection of these F_α is non-empty by assumption. Thus $\{F_\alpha\}_{\alpha \in \Gamma}$ forms a base for a filter which has an adherence point x (Theorem 5.9). Thus $x \in F_\alpha$ for each α, and x is the required common zero. Conversely assume $\mathscr{F}$ is a filter. By complete regularity, for each $x \in X$ and $F \in \mathscr{F}$ such that $x \notin \mathrm{Cl}\, F$ choose a continuous function $f_{x,F}$ such that $f_{x,F} = 0$ throughout F and $f_{x,F}(x) = 1$. Since $\mathscr{F}$ is a filter we get a family of functions satisfying the hypotheses. Let x be a common zero. If $F \in \mathscr{F}$ and $x \notin \mathrm{Cl}\, F$, then $f_{x,F}(x) \neq 0$ which is a contradiction. Hence x is an adherence point of $\mathscr{F}$, so X is compact.

11.3.   Assume that F is closed in X and $x \notin F$. By regularity there exists an open set U such that $x \in U \subset \mathrm{Cl}\, U \subset X - F$. Thus $\mathrm{Cl}\, U$ and F are disjoint closed sets. By Urysohn's lemma there is a continuous $f \colon X \to I$ such that $f = 0$ throughout $\mathrm{Cl}\, U$ (in particular $f(x) = 0$) and $f = 1$ throughout F.

11.4.   Using the hint twice we get finite open coverings of X $(V_1, \ldots, V_n)$ and

$(W_1, \ldots, W_n)$ such that $\mathrm{Cl}\ W_i \subset V_i$ and $\mathrm{Cl}\ V_i \subset U_i$ for each i. By Urysohn's lemma, for each i choose a continuous $f_i \colon X \to I$ such that $f_i = 1$ throughout $\mathrm{Cl}\ W_i$ and $f_i = 0$ throughout $X - V_i$. Let $F = \sum_{i=1}^{n} f_i$, then F is continuous and $F(x) \neq 0$ for all $x \in X$. Now let $\phi_i = f_i / F$. The result follows easily. Finally, to prove the hint, let $A_1 = X -$
$(\bigcup_{i=2}^{n} U_i) \subset U_1$ so by normality there is an open V_1 such that $A_1 \subset V_1 \subset \mathrm{Cl}\ V_1 \subset U_1$. Assume inductively that $V_1, \ldots, V_k$ have been constructed such that $U_i - \bigcup_{j=i+1}^{n} U_j \subset V_i$ and $\mathrm{Cl}\ V_i \subset U_i$ for all $i \leq k$. Let $A_{k+1} = X - ((\bigcup_{i=1}^{k} V_i) \cup (\bigcup_{i=k+2}^{n} U_i)) \subset U_{k+1}$. Then by normality there is an open V_{k+1} such that $A_{k+1} \subset V_{k+1} \subset \mathrm{Cl}\ V_{k+1} \subset U_{k+1}$. If $x \in X$ then $x \in U_j$ for some biggest j. It then follows that $x \in V_k$ for some $k \leq j$.

11.5. We may write ϕ as $\phi(a) = (\phi_1(a), \ldots, \phi_{n+1}(a))$ where each $\phi_i \colon A \to I$ is continuous. By Tietze's theorem we extend ϕ_i to $\Phi_i \colon X \to I$. Define $\alpha \colon X \to \mathbb{R}$ by $\alpha(x) = \sum_{i=1}^{n+1} (\Phi_i(x))^2$. Since α is continuous, $U = \alpha^{-1}(\mathbb{R} - \{0\})$ is open and also contains A. Define $\psi \colon U \to S^n$ by $\psi(x) = \frac{1}{\sqrt{\alpha(x)}}(\Phi_1(x), \ldots, \Phi_{n+1}(x))$.

11.6. Suppose $x \neq y \in X$. If $x, y \in D$ then they can be separated by euclidean open sets. If $x = (x_1, x_2) \in D$ and $y \notin D$, then choose n such that $\frac{2}{n} < x_2$. Then the ball of radius $1/n$ about x has empty intersection with whichever of U_n or V_n contains y. If $x = (0,0)$ and $y = (1,0)$ then clearly $U_m \cap V_n = \varnothing$ for all m, n. Thus X is Hausdorff.

   If A, B are closed neighbourhoods of $(0,0)$ and $(1,0)$ respectively, then there are $U_m \subset A$ and $V_n \subset B$. Let $p = \max(n, m)$, so that $U_p \subset A$ and $V_p \subset B$. Then $(\frac{1}{2}, \frac{1}{2p}) \in \mathrm{Cl}\ U_p \cap \mathrm{Cl}\ V_p$ and so A and B are not disjoint.

Chapter 12

12.1. Let $x_n = 1! + 2! + \cdots + n!$. Then for each n, $(x_i, x_j) \in D_n$ for all $i, j \geq n$, so the sequence $\langle x_n \rangle$ is Cauchy. It is not convergent and therefore the uniformity is not even sequentially complete. Assume for a contradiction that $x_n \to x \in \mathbb{Z}$, then there is an N such that $x_n \in D_6[x]$ for all $n > N$. Since terms beyond x_3 are odd multiples of 3, so is x, that is $x = 6p + 3 = p3! + x_2$ for some integer p. Inductively assume that $x = p_n n! + x_{n-1}$ for some integer p_n. By convergence, for large N, $x_N - x$ is divisible by $(n+1)!$ and therefore $n!(1 - p_n)$ is also divisible by $(n+1)!$. Consequently $x = p_{n+1}(n+1)! + x_n$ for some integer p_{n+1}. Now one can show that for all n, $(n+1)! \geq 2x_n$ (hint:

$1 + 2 + \cdots + n = n(n+1)/2$. Thus the smallest possible value of $|x|$ occurs either when $p_{n+1} = 0$ or -1 and we deduce that for all n, $|x| \geq x_n$ which is a contradiction because $x_n \to \infty$ (in the usual sense).

12.2.   Since $\mathrm{Cl}\, \pi_x \Phi$ is complete for each x, $\prod_{x \in X} \mathrm{Cl}\, \pi_x \Phi$ is complete (by Proposition 12.5). Since Φ is closed, it is closed in $\prod_{x \in X} \mathrm{Cl}\, \pi_x \Phi$ and the result follows from Proposition 12.2.

12.3.   Given $\varepsilon > 0$ it is easy to check that $U_{\frac{\varepsilon}{k}}(x) \subset \phi^{-1} U_\varepsilon(\phi(x))$, so ϕ is continuous. Fix $x_0 \in X$ and inductively define $x_{n+1} = \phi(x_n)$. For $n \geq 1$, $\rho(x_n, x_{n-1}) \leq k\rho(x_{n-1}, x_{n-2}) \leq \cdots \leq k^{n-1}\rho(x_1, x_0)$. For $n > m$,

$$\rho(x_n, x_m) \leq \rho(x_n, x_{n-1}) + \cdots + \rho(x_{m+1}, x_m)$$
$$\leq k^{n-1}\rho(x_1, x_0) + \cdots + k^m \rho(x_1, x_0)$$
$$= \frac{k^m(1 - k^{n-m})}{1 - k}\rho(x_1, x_0) < \frac{k^m}{1 - k}\rho(x_1, x_0).$$

Since $k < 1$, $\langle x_n \rangle$ is Cauchy and hence convergent with $x_n \to x$ say. Since ϕ is continuous $\phi(x_n) \to \phi(x)$, but $\phi(x_n) = x_{n+1}$ and so, by uniqueness of limits, $\phi(x) = x$. If y is another fixed point, then $\rho(x, y) = \rho(\phi(x), \phi(y)) \leq k\rho(x, y)$ which, since $k < 1$, is a contradiction unless $\rho(x, y) = 0$, hence $x = y$.

12.4.   It is enough to prove that if $D = \prod D_j$, where each D_j is an entourage on X_j and $D_j = X_j \times X_j$ for all but finitely many j, then $\prod D_j^* = (\prod D_j)^*$. So, if $(\mathscr{F}, \mathscr{G}) \in (\prod D_j)^*$, then $\mathscr{F}, \mathscr{G}$ are filters on X with a $\prod D_j$-small member in common, that is there is an $M \in \mathscr{F} \cap \mathscr{G}$ such that $M \times M \subset \prod D_j$. So $\pi_j M \times \pi_j M \subset D_j$ for each j and thus $\pi_j M$ is a D_j-small element of $\pi_{j*}\mathscr{F}$ and of $\pi_{j*}\mathscr{G}$. Hence $(\pi_{j*}\mathscr{F}, \pi_{j*}\mathscr{G}) \in D_j^*$ for each j and therefore $(\mathscr{F}, \mathscr{G}) \in \prod D_j^*$. The converse is similar.

12.5.   Assume V is complete in the right uniformity and let $\mathscr{F}$ be a Cauchy filter on G with the right uniformity. Then there is $M \in \mathscr{F}$ such that $M \times M \subset R_V$ the right relation determined by V. If $x \in M$ then $M \subset Vx$, so the trace of $\mathscr{F}$ on Vx gives a Cauchy filter $\mathscr{G}$ on Vx. Since Vx is uniformly equivalent to V, it is complete and thus $\mathscr{G} \to x_0 \in Vx$. Hence x_0 is an adherence point of $\mathscr{F}$ and therefore a limit point of $\mathscr{F}$ by Proposition 8.27.

12.6.   Let V be a compact neighbourhood of e. Then V is complete by Proposition 12.18 and thus G is complete by the previous exercise. The rationals are a non-complete topological group.

12.7.   Let $U_\varepsilon(x)$ denote the euclidean open ε-ball about x and let $V_\varepsilon(x)$ denote the open ε-ball about x in the metric ρ. Consider $\phi \colon \mathbb{R} \to (-1, 1)$ given by $\phi(x) = \frac{x}{1+|x|}$. This is continuous and strictly increasing and so has a continuous inverse. By continuity of ϕ, for each $\varepsilon > 0$ there is a $\delta > 0$ such that $\phi(U_\delta(x)) \subset U_\varepsilon(\phi(x))$, that is $U_\delta(x) \subset V_\varepsilon(x)$. Similarly, using the continuity of ϕ^{-1}, we can show that for each $\varepsilon > 0$ there is a $\delta > 0$ such that $V_\delta(x) \subset U_\varepsilon(x)$. Thus ρ is equivalent to the euclidean metric. However with respect to the metric ρ the sequence of positive integers is a Cauchy sequence which is not convergent (equivalent metrics share the same convergent sequences). To see this sequence is Cauchy:

$$\rho(n, m) = \left| \frac{n - m}{(1 + n)(1 + m)} \right|$$

$$\leq \left| \frac{n}{(1 + n)(1 + m)} \right| + \left| \frac{m}{(1 + n)(1 + m)} \right| \to 0 \text{ as } n, m \to \infty.$$

12.8.   Define $\phi \colon [0, 1) \to [1, \infty)$ by $\phi(x) = \frac{1}{1-x}$ which is continuous and strictly increasing and so has a continuous inverse. As in the previous exercise ρ induces the euclidean topology on $[0, 1)$. Now $\rho(x, y) \geq |x - y|$, so if $\langle x_n \rangle$ is Cauchy in the ρ metric, it is Cauchy in the euclidean metric and thus converges in $[0, 1]$. If $x_n \to 1$ then fix n and let $m \to \infty$, then $\rho(x_n, x_m) \to \infty$ and so the sequence cannot be Cauchy with respect to ρ. Thus $\langle x_n \rangle$ converges in $[0, 1)$.

12.9.   Let $\langle x_n \rangle$ be a Cauchy sequence. Then $A = \{x_n \colon n \in \mathbb{N}\}$ is bounded and so Cl A is also bounded (page 71). By assumption Cl A is compact and hence complete and so $\langle x_n \rangle$ has a limit point in Cl A and thus in X. So X is sequentially complete and thus complete by Proposition 12.10.

12.10. Let $D_\alpha = \Delta \mathbb{R} \cup ((\alpha, \infty) \times (\alpha, \infty))$. If $x \in \mathbb{R}$ and $\alpha > x$ then $D_\alpha[x] = \{x\}$. So the uniformity induces the discrete topology on $\mathbb{R}$ and therefore the closed compact subsets are precisely the finite subsets. As in Exercise 5.4, construct the compact topological space $\mathbb{R}^+$. It is easy to check that $\mathbb{R}^+$ is Hausdorff and contains $\mathbb{R}$ as a dense subset. By Proposition 8.20 $\mathbb{R}^+$ is uniformizable and since it is compact it is a complete uniform space. By Corollary 12.16 it is the separated uniform completion.

Index